Falsification of Special Relativity and the UniKEF Alternative

Dan Keith McCoin

authorHOUSE®

AuthorHouse™
1663 Liberty Drive
Bloomington, IN 47403
www.authorhouse.com
Phone: 1-800-839-8640

First published by AuthorHouse 1/12/2010

ISBN: 978-1-4490-5642-1 (e)
ISBN: 978-1-4490-5643-8 (sc)
ISBN: 978-1-4490-5644-5 (hc)

Printed in the United States of America
Bloomington, Indiana

This book is printed on acid-free paper.

Dedication

This book is dedicated to my wife, Mrs. Linda Lou McCoin, for dedicating her life to me for the past 50 years or to be more honest, "For putting up with me for all that time".

CONTENTS

PREFACE

My journey to the writing of this book began in November 1954, prompted in part by fate when a 15# iron-nickel meteorite landed in our farm yard.

Prior to that time, in my youth I was awed by relativity and idolized Albert Einstein.

However, when I began to study science and physics, I soon realized that some things were not adequately understood.

Newtonian gravity for example, as taught, was strictly a mathematical expression without a physical cause. If gravity were really a force based on mass it made no sense that gravity should be a function of mass squared. It would be linear to the quantity of mass.

That is if I double mass I should get twice the force but empirically if you double the mass you get four times the force. After all two apples did not make a squared apple.

So I began my search for truth and developed views I entitled UniKEF for "Universal Kinetic Energy Field". While initially my objective was to provide a logical physical cause for gravity that matched empirical data, it developed in such a way that it also began to encompass some aspects of relativity.

It grew into a 300 page manuscript with one chapter of calculus on gravity and even made several unique 'Priori Predictions" which over the decades many have become accepted as fact. However, content never became formalized mathematically or reached a point I felt warranted publication. A few things now make it imperative I publish.

1 - I initially concluded UniKEF gravity by physically drawing circles different distances apart and by pencil projecting lines at 1° angle increments through both circles. I then measured each line, and multiplied its length by both the trigonometric function and the circumference of a circle that it would form projected onto a spherical surface the size of the circle being studied by rotating the line through 360° in the Z plane.

I was thrilled when the manual integration of those sums turned out to have an inverse square relationship to the distance of separation 'r'. I had the correct geometric relationship for an energy-based origin of gravity.

An acquaintance, Dr Edward Allard, physicist, also found that result of interest and he invested considerable time and effort to looking at the process using calculus. His results are included here in Chapter 7.

2 - I have conceived an issue that falsifies Einstein's "Special Theory of Relativity". A point that is so simple that it must be published for all to consider.

3 - I have now developed stage 4 lung cancer and the future is unclear. UniKEF is far from complete. It is based on an absolute view where the CMB is an energetic rest reference. I hope by this publication that somebody may pursue the concept and formalize a replacement theory for Special Relativity.

The bulk of this book is my own UniKEF ideas. I take the unusual step of starting the book with Chapter 1 being the CONCLUSION: FALSIFICATION OF SPECIAL RELATIVITY.

Why would I start with the conclusion?

This book is not intended to teach relativity but to expose its shortcomings. I do not go into detail about history, formula derivations or many intriguing aspects of the theory that are not directly involved in the falsification. For having falsified the theory, it makes no sense to try to cover and teach the falsified theoretical details.

While I have had formal education in mechanical, electrical and nuclear engineering; plus post graduate electronics including physics and calculus. I have not used calculus in 45 years and I am not a professional physicist and therefore not qualified to challenge many of the higher mathematics involved in the formalization of the theory.

However, the problem with Special Relativity is not mathematical. It has mathematical utility as is. The problem is the formulation is based on false physics assumptions. Specifically the assumption about spatial length (distance) contraction with relative velocity. I am qualified to discount such mathematics based on flawed physics assumptions used in them. In addition, I wanted to keep this book readable by everyone and avoid overly complex arguments.

I want it made clear up front that UniKEF is not being presented as a replacement theory. It is incomplete, geared more toward gravity, not relativity, and merely offers appropriate questions about relativity and hints at some possible solutions

Professionals that might actually start to read this book would trash it before completing the first chapter if every detail were not spelled out with formula derivation, diagrams and lengthy explanations.

They would just conclude that the writer did not know the subject. However, by its title I hope the book will cause many relativists to take the bait, at least crack the cover and find that they cannot answer the challenge when they are confronted with the truth. If they actually give the issue proper, deference progress may be made.

For non-professionals, Chapter 1 CONCLUSIONS, will appear to be so obvious they will wonder why the book has been written at all. That is the irony of this falsification, it is so obviously simple, yet for 104 years, modern science has been so blinded by the "Counter Intuitive", nature of Special Relativity and the fact that mathematically it has utility, that they either missed or deliberately ignored the obvious errors in basic physics.

A case of not seeing the forest for the trees is an appropriate analogy.

When Einstein published his theories of relativity, they were not well received. But through a series of predictions such the bending of light around a star and correctly predicting planet Mercury's orbit (which Newton's formula doesn't do) it resulted in a dramatic interest and ultimate acceptance of Einstein's views in spite of some rather bizarre predictions. The theory now has status of God like law in modern physics.

Acceptance of Special Relativity was accompanied by an unwarranted abandonment of common sense and basic physics. Normal standards were replaced by the phrase "It is Counter Intuitive" which is an escape goat from reality because the truth is that much of it is physically impossible and nonsense.

While this book contains a falsification claim against Special Relativity, I support the principle of relativity. Relativity is indicated by empirical data but it is not Einstein's relativity that is physically real.

The primary error of modern science is to see all test data as proof of Special Relativity when such information may support the Principle of Relativity but is not exclusive results for Special Relativity as a cause.

Such test results are proof only of a particular relationship and not of its cause, be it Special Relativity or some alternative physics.

A perfect example is the argument that cosmic muon flux at earth's sea level proves Einstein's theory. The reason is that muons have a statistical life span of around 2.2μ seconds (millionths of a second) and at the speed of light, they could only travel 660 meters.

A calculation using Lorentz formulas, in the theory of relativity gamma factor, predicts time dilation that accounts for the approximate flux level of muons actually reaching the earth's surface, and hence is praised as proving Einstein's view.

The truth is that there are other explanations, which involve a different relativity. The falsification of the muon Special Relativity claim is discussed in Chapter 3 "Fallacies".

Another thing that is ignored is that 1,000,000 correct predictions can be falsified by merely one incorrect prediction. The primary argument put forth by most relativists is "It has been successfully tested for over 100 years and there has been no test data that contradicts Special Relativity's predictions.

That argument is not entirely correct or as a minimum is misleading. There have been hundreds of tests, paper studies, etc., that claim failure of Special Relativity but they are not replicated to confirm or falsify the claim, they are shunned by mainstream journals, ignored or attacked by mainstream relativists without any due diligence being applied.

The fact that it is claimed no test data disproves Special Relativity is based on two points:

1 - Relativists ignore any and all such test data and ridicule the authors without any due diligence to confirm or falsify the claim.

2 - Special Relativity is subjective involving thought experiments that appear impossible to test in real time. Most of Special Relativity theory has never been tested.

Most relativists will tell you that the only thing important is that the theory allows you to make valid predictions. They do not care that some predictions are not verified by test or that the concept is physically flawed in areas not used.

Unfortunately, that means they are not looking at alternatives the way they should but look at all tests they accept in the guise of proving Special Relativity, not just supporting the principle of relativity.

There have been no tests of the "Counter Intuitive" reciprocity aspects of Special Relativity. Reciprocity of time dilation appears impossible to test. If so Special Relativity fails to meet the standard for valid theory.

The scientific standard for a theory to be valid is that it can be tested.

Anyone noting these facts is accused of advocating a conspiracy and that thousands of scientists would have to be guilty of conspiring to defraud. There is no conspiracy the issue is driven by ignorance or arrogance. The continued survival of Special Relativity is advanced by fear of ridicule by others, of job or funding loss, if you question relativity.

I repeat some issues in this book a number of times because it is important that they be understood and not missed.

CHAPTER 1

CONCLUSION: FALSIFICATION OF SPECIAL RELATIVITY

If you do not know Special Relativity, the Conclusion will appear to be a most obvious statement and you will wonder just why it has been published and has any significance. You will have to read Chapter 2 & 3 to find out.

If you know Special Relativity, you may appreciate the simplicity of the falsification. Then again, my experience has been that appreciation is not a proper adjective. You will feel rebuffed and not give the issue proper deference.

In case you do feel rebuffed, you will likely not read further. Although I would hope that, your interest in my views might have become piqued and you will consider the UniKEF concept.

BASES: Physically real is defined as affects that are permanent and exist subsequent to having had relative velocity and not some "Illusion of Motion" measurement made during relative velocity that is observer dependant.

Postulates:

1 -You must have a physical cause for a physical affect.

2 - Anything physical must be physical in all frames, and not be observer frame dependant.

3 - Measured invariance of light is an illusion of its production or measurement and not of its propagation.

There is empirical data supporting time dilation as being physically real. There is no empirical data supporting spatial length contraction as being physically real.

Given these facts and postulates, the following diagrams set the stage for the Conclusion.

PROOF: Hypothetically using v = 0.866c where γ = 1/(1 - $v^2/c^2)^{1/2}$ = 2.0 and times are in hours.

CASE 1: A time dilated clock.

A round trip from point "A" to "B" and back to "A"

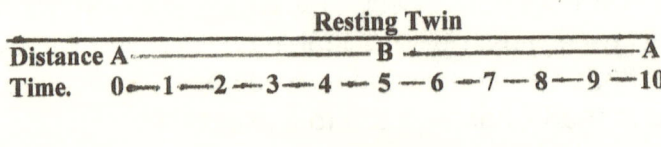

This scenario matches Special Relativity and empirical data. A traveling twin would be younger upon return home.

CASE 2: Length (Spatial Distance) Contraction. Same γ and speed but - v.

A round trip from point "A" to "B" and back to "A"

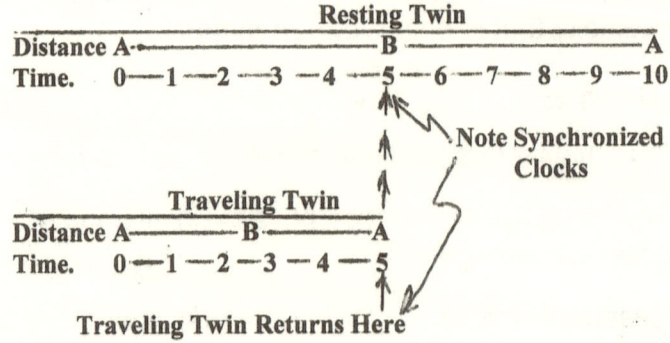

Predictions of Special Relativity cannot be met because no time dilation has occurred.

Both the traveling clock and resting clock remain in synchronization as shown by the diagram and must display the same time upon the twins return.

CONCLUSION:

Special Relativity is falsified because it stipulates the traveling twin moving at 0.866c will go 1/2 the distance, in 1/2 the accumulated time, at a common measured velocity to the resting twin and therefore returns home younger than his brother.

The Case 2 diagram proves clocks under those conditions do not change tick rate and are not dilated. Spatial length (distance) contraction is not a real physical affect and does not cause or allow for differential aging as claimed by Special Relativity.

Assumptions made to include spatial distance contraction in the mathematical process are formalized and then when the mathematics is used relativists claim, "See" it exists". However, that is circular logic because it is assumed in the mathematical formulation at the outset.

Where TT is Traveling Twin and RT is Resting Twin.

.....____TT___ _RT_

v = 0.5d / 0.5t = d / t

In other words should the traveling twin continue to go the full distance he will accumulate the same amount of time as the resting twin. Do not be confused by "Full Distance" versus "Round Trip".

The Special Relativity argument is that the Round Trip is 1/2 the distance for TT as it is for RT. Such that the TT Returns home in 5 hours, not the 10 hours according to RT's clock. By

3

Special Relativity that makes him arrive back younger than his stay at home twin.

However, the reality is that no time dilation can occur. The resting clock and traveling clock must tick in unison and display the same accumulated time. Einstein constructed Special Relativity using false assertions backed by mathematics contrived to support the assertion.

Put into everyday terms assume it is 60 miles between city "A" and city "B" measured at rest to the cities. The velocity I use in this example will be 60 Mph but assumes the affect of going 0.866c, which according to Special Relativity, time and distance will be cut in half for TT.

According to RT, TT will go 60 miles in 1 hour or 3,600 clock ticks of 1 second each.

That is 3,600 ticks/60 miles equals 60 ticks/mile.

Now according to Special Relativity TT only goes 30 miles making the same trip and accumulates a mere 30 minutes. It is claimed therefore that clocks have become dilated since at the end of the trip TT has accumulated one half the time and he therefore is now younger than his stay at home twin.

However, since TT took 30 minutes to go 30 miles, tick rate equals 1,800 seconds / 30 miles = 60 ticks/mile. Both clocks are ticking in synchronization they are not tick dilated and Special Relativity is internally inconsistent in its physical assertions.

Spatial length (distance) contraction does not result in or allow for time dilation. Time dilation only occurs if clock tick rate is physically changed by an actual change in velocity.

"Actual Velocity" and "Absolute Velocity" are prohibited terms in Special Relativity. All motion is relative.

Since clocks are ticking in unison in Case 2, RT must display the same accumulated time of TT when TT returns home. That precludes RT from seeing TT completing the trip according to the stipulation by Special Relativity that RT's clock recorded 1 hour when TT returned.

Special Relativity is physically and rhetorically internally inconsistent on detailed analysis.

While velocity is symmetrical in absolute terms and all third parties in all frames, see relative velocity as being symmetrical. Nevertheless, at the local level computing velocity using different dilated conditions of clocks will produce different calculated velocities.

Where rest distance is 60 miles, tRT = 1 hour and tTT = 30 minutes.

vTT1 = d / tTT = 60 miles / 0.5 hour = 120 Mph - Mandated by physics.

vTT2 = d / tTT = 30 miles / 0.5 hour = 60 Mph - Assumed by Special Relativity by using common tick rate standards and symmetrical relative velocity.

vRT = d / tRT = 60 miles / 1 hour = 60 Mph - According to RT and everyday physics.

The assumption in Special Relativity that relative velocity remains symmetrical at the local proper calculation, forces the time dilation between frames to declare distance has changed. Special Relativity disregards the prior stipulation that in the rest frame the traveling clock is dilated (ticking slower).

An analogy of time dilation in Special Relativity is to consider Bill and Bob going from town "A" to visit John in town "B". The measured distance between cities is 60 miles apart. Bill and Bob drive separate cars. Bill's car and watch are working fine.

However, Bob's speed-o-meter is broken and since there are radar devices all along the route Bob tells Bill "You set the pace at 60 Mph and not a bit over because I cannot afford to get another ticket".

Bill agrees and they call John to let him know they are on the way. John assumes they will drive the speed limit and expects to see them in one hour. However, Bob does not know the batteries in his watch are low and it is only ticking 8 times for every 10 ticks of Bill and John's watches.

Under those conditions, Bob's watch will only accumulate 48 minutes for the trip.

At the end of the trip, John tells Bill "You made good time" and Bill says, "Yes we were on cruise and drove a steady speed all the way". Bob looks at his watch grabs his calculator and computes: 60 miles / 48 minutes = 75 Mph!

Bob yells at Bill, "I told you I couldn't afford another ticket and you were driving 75 Mph. I could lose my license".

Bill and John argue with Bob about the time because Bill swears he drove the speed limit of 60 Mph and John and Bill agree it took just one hour. Finally, they conclude that Bob's watch must be running slow.

They never consider that Bob drove less distance to get there because Bill and Bob drove side by side all the way and his odometer registered 60 miles just as the mile markers indicated. Bob's watch was slow in both frames of reference just as logic for anything physically real mandates.

Einstein's length contraction doesn't allow for the stipulated time dilation affect he claimed and spatial distance contraction is an irrational physical result that is unsupported by empirical data leading to many absurd consequences not generally realized even by professional physicists.

For example if you follow the mathematics to their ultimate conclusion you will find that according to Special Relativity accelerating away from earth you will reach a point mathematically where the faster you recede (move away from the earth) the closer you will get to it!

The formula for distance in Special Relativity is "Distance Measured at Rest" is d and "Distance Measured by a Moving Observer" is d'. d and d' have the following relationship where v is velocity of the moving observer and c is the speed of light.

If you assume you are a pilot of a rocket passing our moons orbit 240,000 miles away from Earth at 0.866c then according to relativity from your frame of reference you are:

$$d' = d (1 - v^2/c^2)^{1/2} = d(1 - 0.866^2)^{1/2} = d (1 - 0.75)^{1/2} = d (0.25)^{1/2}$$
$$= d * 0.5 = 240.000 \text{ miles} * 0.5 = 120,000 \text{ miles from Earth.}$$

If you now go full throttle and accelerate to 0.9c you will have a distance of:

$$d' = d(1 - 0.9^2/c^2)^{1/2} = d (1 - 0.81)^{1/2} = d (0.19)^{1/2} = d * 0.435889$$
$$= 104,613 \text{ miles.}$$

So instantly upon your change in velocity AWAY from Earth you got 15,386 miles CLOSER! To emphasize this lunacy even more consider not accelerating but firing a missile from the rocket away from you and Earth.

I'll ignore the velocity addition formula for the moment since it does not alter the ultimate out come. The missile is now supposedly closer to Earth than you and traveling faster. You have just shot yourself in the back!

This is unacceptable physics.

Also if you follow the mathematics you will learn that for a particle accelerating from zero velocity in the lab to 99% the speed of light in 10 micro-seconds means that the distance to the edge of the universe has contracted to 14% of it's original distance in that short time interval.

Taking the general radial distance of the universe as 15 billion light years, it will reduce to 2.1 billion light years in

0.00001 seconds or a distance change rate of 4E13 times the speed of light. That is 40,000,000,000,000 (40 trillion) times faster than light.

When I was in one of several formal debates with a particle physicist and raised this issue his reply, after a few days, was to produce some mathematics which claimed that when this occurs it happens behind an event horizon and therefore was OK. In other words, he did not argue that it does not occur but rather that you would not see it occur; and therefore the FTL function did not violate Special Relativity.

Not seeing something happen is not the same thing as it not happening and hence the argument is not persuasive. If your loved one is in a serious car crash and you aren't there to see it, doesn't alter the fact that the crash occurs.

If by collapsing the dimension of the universe in this fashion I should happen to collide with a massive body I suggest the "I didn't see it coming" does not hold much merit in physics either.

Special Relativity prohibits travel faster than light speed through space but causes velocities of objects to move spatially at velocities that far exceed the speed of light.

It is claimed that expansion and contraction of space doesn't count against Special Relativity that only motion through local space is limited to $v = c$. It seems arbitrary or selective to say that expanding or contracting space are not applicable to Special Relativity but then have Special Relativity cause expanding or contracting space.

Lastly, if you accept Special Relativity then you must also accept that the current observation that the universe is undergoing an accelerating expansion is in fact just the opposite. Objects near the edge of the universe and moving near light speed, which appear to be undergoing an accelerating expansion, must actually be undergoing an anti-Lorentz Contraction affect by decelerating.

Absolute energy based on absolute universal velocity is a logical cause of clock tick rate dilation. However, "Absolute Velocity" is prohibited physics in Special Relativity by fiat, not fact.

What is now "Counter Intuitive" is that at luminal speeds relative velocity will no longer compute the same according to a traveling observer versus the resting observer.

At day-to-day velocities, it is logical that if I say, you are going 30 Mph faster than me; you would say I am going 30 mph slower than you are. That is the common sense everyday symmetry of relative velocity.

However, once you reach luminal speeds and clocks begin to dilate then locally calculated speed will not be symmetrical. Velocity is a computed value based on the ratio of two physical parameters, time and distance by the formula:

$v = \Delta d \, / \, \Delta t$

When d is known, as in the examples given, and clocks accumulate time at different rates then each observer will compute a different velocity. Universally relative velocity is still symmetrical but we have no universal reference by which to compute the velocities.

This issue opens up an argument against the velocity addition formula used in Special Relativity and falsifies the length contraction component of Special Relativity but adds the "Counter Intuitive" fact that locally computed velocity is not symmetrical at relativistic speeds.

At every day speeds clocks remain in useable synchronization even though we know from GPS that is not physically correct. There are minor changes in tick rates that GPS must correct to function properly.

At an orbital speed of around 2.4 miles/second, (8,665 Mph) clocks dilate and lose 7.2 micro-seconds/day due to orbit velocity. That is a clock error of 1 second in 380.5 years.

However, gravity has a greater inverse affect than velocity such that General Relativity causes orbit tick rates to increase by 45 micro-seconds/day. There is therefore a net +38 micro-seconds/day increase in orbit tick rate.

GPS Clock time dilation due to velocity matches empirical data and confirms the principle of relativity.

Clock time dilation preserves relativity but without length contraction of space.

Einstein's Theory of Special Relativity is falsified by the basic formula $v = 0.5d / 0.5t$. This simple formula re-states the assertion by Special Relativity that the traveling twin goes one-half the distance in one-half the amount of time at a common relative velocity of 0.866c.

However, where $d = 1$ and $t = 1$ it also shows that:

.....____TT___ _RT_

$v = 0.5d / 0.5t = d / t = 1.000$

That means that the resting clock and traveling clock both tick in synchronization hence there is no time dilation as advocated by Special Relativity during the purported length contraction.

As such, the resting twin's clock must display the same accumulated amount of time upon the traveling twin's return. It is not possible therefore for the resting twin's clock to accumulate the full time asserted by Special Relativity. Unless the traveling clock physically dilates as in case #1.

The assertion by Special Relativity that the traveling twin returns home younger due to going less distance via Lorentz

Transforms and the space cone mathematics is Falsified and a different relativity is mandated because time dilation is empirically indicated.

However, if you physically dilate a clock then spatial length contraction is prohibited since that compounds any relativistic affects and accumulated time will no longer match empirical data.

CHAPTER 2

The Fundamental Relativity Issues

Special Relativity is a highly mathematical construct where time and space merge into a flowing relationship due to relative motion. One transitions from distance to time or time into distance. Collectively however everyone agrees on the time-space interval. In Special Relativity "Time and Space" become "Time-Space", one entity with two key attributes.

Special Relativity asserts that a common relative velocity causes a clock to slow down in one frame of reference while causing distance to contract in another frame of reference. It is not rational to hold that a common cause produces two different physical affects based on an observer's perception.

Producing two different physical affects from one cause that is frame observer dependant is not to claim the theory fails mathematically.

The mathematics of Special Relativity is based on the assumptions of the time-space format but the length contraction component is only applied mathematically and is never of real concern physically. It is just ignored, it only provides mathematical continuity and utility to model the concept to Einstein's desires.

Recall that according to the theory either one of the observers can declare to be at rest and making it the other observer that has the velocity. Therefore, Special Relativity is dealing with perceptions while in motion or "Illusions of Motion" measurements, not physical changes.

Time does dilate; clocks compared in a common rest frame subsequent to having had relative velocity will have accumulated different amounts of time. Only one of the two frames experience the dilation and since both share the same relative velocity it proves that relative velocity was not the cause of the differential.

Not all observers having experienced the relative velocity actually display any permanent physical change. Only a frame that has accelerated to switch frames ultimately displays a loss of accumulated time. I repeat, "Since both experience the same relative velocity and only one is affected it proves that mere relative velocity is not the cause of permanent changes."

The velocity of light in a vacuum (space) is just under 3E8 m/s. That is 300,000 Kilometers/second or 186,282 Miles/second. The small letter 'c' represents the speed of light. It took Apollo 11 three days and two hours to reach moon orbit (76 hours) and light requires just over 1 second.

An inertial frame is a condition where you are moving at a constant velocity with no shift in direction or acceleration. In such a frame, a person cannot tell if he is in motion or at rest and others are moving. Similar to flying in a commercial jet liner at 350 Mph and watching the earth and clouds move by you. According to Special Relativity any inertial velocity is considered being at rest.

The assertion by Einstein that given two inertial frames with relative velocity, either can assume to be at rest and it is the other that has all velocity, on the surface seems logical.

However, in Einstein's Special Relativity if two inertial frames have a relative velocity to each other and one assumes to be at rest then the other undergoes certain relativistic changes in time or distance.

If the relative velocity is 0.6c then the frame considered in motion is moving at 111,769 miles/second. If according to my watch at rest 1 hour has passed then you traveled 402,369,120 miles in my frame but your watch will have only accumulated:

$$t' = t (1 - v^2 / c^2)^{1/2} = 60 \text{ Min} \times 0.8 = 48 \text{ minutes.}$$

Further, in your frame since relative velocity is assumed symmetrical, and in your frame, your watch is ticking normally, you must see distance contract because you flew for 48 minutes at 0.6c such that you only traveled:

$$d = vt = 6,706,152 \text{ miles/minute} \times 48 \text{ minutes} = 321,895,296$$
miles.

Therefore, going from the same point "A" to point "B" you traveled 80,473,824 fewer miles than I say you did as is measured between points when at rest.

However, this only occurs because Special Relativity did not carry the stipulated dilated clock condition of the resting frame into the moving frame. They reset the time standard by claiming 1 second in the moving frame is still the same as 1 second in the resting frame, after having first declared it was dilated; a clever mathematical bait and switch.

While you may not measure or know your time has slowed down because all physics must also slow down, your clock must tick more slowly physically to account for accumulating less time for the trip or according to Special Relativity you must have gone less distance.

A dilated clock fully accounts for trip time ONLY if distance does NOT contract. By claiming, clocks are dilated in one frame but not in others, forces distance to change. Any permanent affect must be physically real and only time dilation or length contraction can be physically real because if both were physically real it would compound the affect and no longer match empirical data.

More importantly, you must understand that the view in Special Relativity has reciprocity, which means my clock is ticking slower than your clock at the same time your clock is ticking slower than my clock. That physical impossibility is treated as merely being counter intuitive.

It is a physical impossibility and must therefore be relegated to being mere perception and not physical reality. Einstein in fact said "Sees" which implies perception and not physical reality. However, many modern day physicists like to claim it is physically real.

The fact is however if you are also traveling in an inertial frame you can equally claim to be at rest and that it was I that had the 0.6c velocity. Within the 48 minutes, I say your watch accumulates for the trip you would say my clock only ticked 0.8 times the amount of your watch so you would expect my clock to have only accumulated 38.4 minutes not the hour initially stated in the problem.

However, Special Relativity does not carry stipulated physical conditions for one frame into the other frame, which is part of the problem of Special Relativity as a physical concept.

It declares a clock time dilated in one frame but then when computing information in the moving clock frame it switches physical time standards and no longer considers the clock dilated which then forces distance to have changed because relative velocity has been declared symmetrical.

Therefore, the results from my frames view will have you suffer the same changes that you thought I had. That is your clock will only accumulate 48 minutes to my clocks 1 hour. If something is physically real then it must translate into physical reality in any frame of reference.

Relativists don't normally put things in this context but assert your clock sees me take 1 hour to traverse the distance and my clock will accumulate 48 minutes and I will only travel 0.8 times as far as you see me travel. However, realistically one should keep both views concurrent to see the true fallacy of the claims. You must remember the lesson of Case #2 where length contraction is shown to have clocks tick in unison.

You have a situation where you are claiming that two clocks both each tick slower than the other does at the same time. This situation is generally referred to as being merely "Counter Intuitive" but in reality is physically impossible.

However, this is a manifestation of modern misinterpretation since what Einstein actually said was they each "See" the other time dilated.

"Seeing" need not reflect physical reality. Measurements made during relative motion are affected by an "Illusion of Motion". It would be equivalent of putting on a pair of red glasses and then insist that the entire universe was red. That of course would not be the reality but is only a "Perception" or an illusion.

"Seeing" each other dilated during relative velocity is therefore a matter of "Perception" and is not physically real. Relative velocity time dilation is caused by an "Illusion of Motion" and vanishes upon termination of the relative velocity. Any physically real affects must persist after relative velocity has terminated.

However, there is an inherent problem with reciprocity solution in the twin paradox.

This purported paradox has been declared solved by what is generally called "Frame Switching". This simply means the traveling twin accelerated and became non-inertial to change velocity while the stay at home twin remained in an inertial frame. This purportedly breaks the symmetry of relative velocity and makes the traveling twin the younger brother.

The Lorentz Transforms and Space Cone mathematics support all this but then again they were designed with the assumptions they use.

17

Of course, they do not explain why such break in the symmetry should affect the result since the result is based on the inertial relative velocity between the two brothers, not with breaking of symmetry. However, it at least has provided cover for the flawed theory and people could ignore the balance of foolishness the theory advocates.

It should be obvious to all that mere "Relative Velocity" does not cause the loss of accumulated time known as time dilation because both brothers share the same relative velocity even during acceleration by the traveling twin. Therefore, something other than relative velocity must be the cause.

Do not let relativists confuse you with their mathematics because the mathematics is merely formal processes based on the assumptions of the formulation.

Do not be confused. I am not claiming acceleration causes time dilation it does not. However, acceleration is required to change velocity. It is the absolute change in velocity (energy) that causes the time dilation.

This can be proven by the following exercise:

Three clocks "A", "B" & "C" are at common rest at location "C" and "A" & "B" are launched out to the ends of a prearranged course where they have equal acceleration range from points 'X' to the test start lines "S" which are equal distance from "C" in opposite directions.

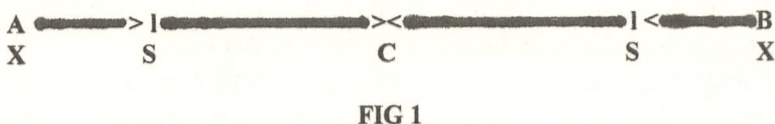

A ⬤━━━> | ⬤━━━━━━━> < ⬤━━━━━━━⬤ | < ⬤━━⬤B
X   S   C   S   X

FIG 1

Once there "A" & "B" send "Ready" light signals to "C". Upon receipt of both signals "C" then simultaneously sends "Start" light signals to both "A" & "B". They immediately upon receipt launch simultaneously according to "C" and accelerate equally until they simultaneously reach the "S" line at v = 0.6c. Both set their clocks to t=0, cut their rockets, go inertial, and send "Confirmation" light signals to "C".

Upon the simultaneous receipt of the "A" & "B" confirmation signals "C" then sets his clock according to the time required for the light signals to travel from the "S" positions based on known distance and the speed of light 'c'.

Assuming "S" is 0.6 light hours from "C" then "C" knows to set his clock to 0.6 times 60 minutes = 36 minutes so that all three clocks will have been set to t=0 at the same time according to "C".

At the 0.6c, velocity "A" & "B" will require 1 hour to make the trip according to "C". However, as "A" & "B pass by C", simultaneously they transmit their clock data to "C" by digital signal and due to time dilation they will have only accumulated:

$t' = t(1 - v^2/c^2)^{1/2} = (1 - 0.6^2)^{1/2} = (1 - 0.36)^{1/2} = 0.64^{1/2} = 0.8 \times 60$ minutes = 48 minutes.

Their clocks have accumulated equal time because "A" & "B" had symmetrical acceleration and equal velocity to the common rest frame. What is important here is that in spite of having had relative velocity to each other "A" & "B" clocks accumulate the same amount of time. Since both clocks, tick in unison relative velocity did not generate time dilation according to the prediction of Special Relativity.

Further according to Special Relativity the velocity between "A" & "B" is not 0.6c + 0.6c = 1.2c because nothing can exceed c. They use the velocity addition formula and insist that relative velocity between them is w = u + v / (1 + vu/c²) = 0.8824c.

Having a relative velocity of 0.8824c mandates that each would see the others clock ticking at 1,000 ticks to their 2,125 ticks. In addition to the different clock tick rates you now must have, different distances traveled as well.

Clearly, these are not physical realities since terms change with observer perception.

Relative velocity did not generate any time dilation between them. They could just as well have launched from the same end of the course and have been co-moving with no relative velocity because the result would be the same. Relative velocity plays no role in time dilation.

The only real physical time dilation is to the common rest frame and is valid between clocks ONLY for the accelerated clock and if the other clock has remained in the common rest frame. The SR view that "A" sees "B" and "C" dilated and vice-versa for "B", does not exist once relative velocity terminates.

The Velocity Addition Formula is designed to insure nothing can ever go faster than the speed of light 'c'.

It states that $w = (v+u)/(1+vu/c^2)$. Where w is the compound velocity.

This concept comes into play if you are on earth watching a rocket flying through space at 'v' = 0.8c and the rocket fires a missile that has a velocity 'u' = 0.4c to the rocket then:

$w = (0.8 + 0.4) / (1 + 0.8*0.4) = 1.2 / (1 + 0.32) = 1.2 / 1.32$ =0.909c

In other words, 2+2=3 and the missile is only moving with a velocity of 0.909c relative to you, not the 1.2c that would be the normal 2+2 = 4 mathematics. While you see the missile moving a mere 0.909c - 0.8c = 0.109c relative to the rocket the rocket sees it moving at 0.4c.

These numbers are not failures of theory per se because they fit theory. The problem is the mathematics has been designed to fit theory using flawed physical principles. The v =

c limit may well be an absolute function but disregards the local calculated velocity based on different proper clock tick rates.

Einstein's errors were:

1 - Asserting locally calculated velocities are symmetrical and not recognize that the symmetry is an absolute condition that he rejected.

2 - Asserting distance has contracted because the moving clock accumulated less time rather than carry forward the physical stipulation that a clock has become dilated and is ticking more slowly; which fully accounts for the accumulated trip time ONLY if distance did not change.

3 - Disregarding the respective equal tick rates of clocks in both frames in the spatial length contraction view.

4 - Not requiring a physical cause for physical affects and having them frame dependant.

There are different versions of the velocity addition formula depending on if objects are approaching or receding away from you. It is a key element of Special Relativity to insure that nothing exceeds $v = c$ in a vacuum.

Where modern science has gone wrong is absolute faith in mathematics. If mathematics predicts something and there is no physical evidence to refute the result, it is considered valid.

The fact is that reciprocity and length contraction have had no empirical data to support their existence but then there have been no tests of the concepts at luminal velocities to try to produce such evidence. Since nothing has falsified it to date then it is considered valid.

It appears in fact that testing reciprocity and length contraction at relativistic velocities is impossible; in which case Special Relativity is falsified at the basic standard for theories. Scientific standards require a theory be testable to be considered valid.

In any case, the situation is that mathematics is nothing more than a formal process and it is based on assumptions and inputs. Mathematics does not know, care or test if those assumptions are valid or false. Consequently, Special Relativity appears to be internally consistent "mathematically" and makes good predictions; hence has utility and is blindly accepted.

Nevertheless, it is flawed at the most basic physics level in the assertion that spatial length contraction is required to account for time dilation or has a consequence of slower aging.

Assuming it does however make the mathematics work.

Modern science has been so enamored and awed by the merged time-space concept and its bizarre consequences, that they have ignored basic physics.

With respect to the viewpoint about ether or an absolute frame of reference the modern day assumption that:

"Absence of Evidence" is "Evidence of Absence"

That mentality is very flawed science.

Put into a different context Special Relativity is based on negatives. That is since there is no empirical evidence of ether and we have not identified an absolute rest frame then they do not exist.

This same scientific standard can also support the argument that "Flying Pink Elephants are invisible and since we do not see them, therefore they must exist".

An Einstein quote that is most appropriate at this time is:

******************************* Quote ***************************

"As far as the laws of mathematics refer to reality, they are not certain, and as far as they are certain, they do not refer to reality."

Albert Einstein

****

Albert Einstein has been idolized as a genius even though he is known to have made errors. He in fact left school without a diploma. He failed an engineering entrance exam in 1895.

In college, his professor Bartocci (and others) alleged that he was bad in mathematics and relied on others for problems that were more complex. Even his wife Mileva Maric Einstein was scientifically minded and shared his interests. She even fared better in physics in school than Einstein. Many have openly asserted that she contributed to his many papers without credit.

She got an overall score of 4.0 and failed, he got an overall score of 4.9 where 5.0 was a passing grade but they rounded his grade up to 5.0 for him to pass.

It is alleged by some that he plagiarized many of his papers from others. While that has not been proven, it is historically clear that others preceded him with many of the basic ideas.

1 - Photon theory of light

2 - Mass to Energy

3 - Brownian motion in liquid

4 - Theory of Relativity

Maxwell published work showing the invariance of light in 1887.

Fitzgerald wrote about material contraction in 1880 and Lorentz published his formulas in 1904.

Poincare was working on the Principle of Relativity in 1898 and actually published a short paper three months before Einstein about relativity. Einstein's contribution being giving it its name "The Theory of Special Relativity".

Thomson & Kaufmann in the 1890's and Poincare in 1900 wrote about mass energy relationships. Olinto De Pretto in 1903 published a paper with the expression $E = mc^2$ as did Hasenohrl a similar version in 1904.

Many modern day scientists' and textbooks teach that Einstein invented the atomic bomb. He did not. The actual truth is that other scientist such as Physicist Leo Szilard and Eugene Wigner did but had no influence with those in power so in July 1939 they explained the problem to Einstein. Einstein replied, "The possibility of a chain reaction has never occurred to me".

However, he quickly understood the concept and implications and signed a letter in August of 1939, prepared by Szilard, to President Franklin D. Roosevelt urging him to build the bomb.

Brown developed the Brownian motion of liquids theory in 1827. He was followed by Gibbs and Boltzmann in the 1890's.

Wein and Planck started quantum theory. By 1900, Planck had developed the concept of quanta of energy. Einstein merely restricted his work to quanta of light energy called photons.

Whatever the truth about Einstein and plagiarism, regarding his lack of recognition of others before him in his papers:

1 - The fact is that he used the 'k' term developed by Brown.

2 - Much of his work contains components with names of other scientists such as the Lorentz-Fitzgerald contraction formulas.

3 - Others actually published the term $E = mc^2$ before him and The Special Theory of Relativity being comparable to

Poincare theory of relativity leaves serious questions as to why Einstein receives as much credit as he does.

It seems very little original thoughts actually came from Einstein but he was quick to compile, apply, name and publish the works of others without giving names like Pretto, Grossman, Reimann, Heisenberg, Lamor, and others any recognition. Hilbert preceded Einstein with General Relativity.

Hilbert received a letter from Einstein 18 November, 1915, thanking him for the advanced copy of his lecture that he then gave about General Relativity on November 20th.

Einstein gave his speech about General Relativity on November 25, 1915. He had received the advanced copy of Hilbert's lecture two weeks before announcing General Relativity without any mention of Hilbert.

Charges of plagiarism were made by Hilbert against Einstein and vice versa however, three Einstein supporters (Cory, Renn & Stachel) published a paper in 1997 claiming Hilbert's lecture did not contain the critical field equations presented by Einstein a week later in his lecture.

However, Professor Winterburg, physicist, at the University of Nevada, Reno, disputed this conclusion in 2003 when he noted that the galley proofs used to clear Einstein of plagiarism of Hilbert's lecture had been tampered with by part of a page having been cut off.

He further noted that the missing information contained the proper field equations as supported by other galley proofs in other forms. Einstein in fact had been invited and spent a several days at Hilbert's home just before this incident. Professor Winterburg found that the prior declaration of innocence made in 1997 was but a crude attempt to change the historical record.

The preceding quote by him, about mathematics and reality, is as though and he was laughing behind our back over his tactics. We can only conclude that he really was a genius or a very shrewd and clever plagiarist.

Special Relativity is based on an arbitrary circle of mathematics and rules that protect it from challenge. Even the obvious falsification I present here will likely be ignored or mitigated by double talk about simultaneity or some other rule of Special Relativity.

The argument will be made for the pure mathematical aspects of the theory claiming Lorentz Transformations account for the discrepancy but the reality is Special Relativity is based on flawed logic supported by mathematics designed to support the flawed assumptions.

Special Relativity is a product of circular logic. Flawed logic and assumptions are used to construct a mathematical treaty, and then the formalized mathematics is used to claim the assumptions are proven.

CHAPTER 3

Fallacies:

Any who dare challenge Special Relativity are labeled Crank, Crackpot or worse. Being labeled a Crackpot is an interesting process. It seems one of the standards by which you become labeled is to deny being a Crackpot and to refuse to convert to their view point by merely repeating your own. Even though relativists continue to repeat their dogma without addressing the physics issues raised against them.

Their response is "But that is not what Special Relativity says". Physics has become a religious belief system and is no longer really physics. Merely repeating the assertions made in Special Relativity does not alter the physical reality flaws..

I am fortunately old enough to be on social security and do not need to worry about being Black Balled by mainstream science. I have no threat of job loss, research-funding loss and sticks and stones will not break my bones.

So label me if you choose for I can label you, indoctrinated, brainwashed, arrogant, egotistical and incapable of independent thought. Until you properly address the physics issues your mathematics have no meaning.

Mathematics is important and necessary in physics. It can help describe reality but currently it is used to try and create reality. It is applied without knowledge of contingencies or factors being known that limit its applicability and bizarre extreme conditions are then advanced as being physically real without any evidence or empirical data to support the conclusion.

Special Relativity is protected to some degree by embarrassment. In that to object to it infers you are not as smart as the person that preaches it as physics gospel. Relativists do not hesitate to invoke that innuendo and that suppresses many scientists from speaking out.

The attitude "I have studied physics and higher mathematics and understand Special Relativity but you don't" or "If you don't agree with Special Relativity then you are showing your ignorance" are prevalent among relativists. Many good scientists are afraid to speak out for that reason.

I have participated in thousands of discussions and find it amusing that relativists will, after a few rounds of debate, claim I cannot or will not learn; that I just keep repeating myself. While they do not realize that, they also are just repeating themselves and are not addressing the issue in terms of physics but are merely quoting theory as proof of the theory.

For those interested there is an organization, the "Natural Philosophy Alliance (NPA)". It is free to join that organization which has over 1,600 members. It is comprised of nonprofessionals, physicists, mathematicians, engineers, scientist of all types that are actively seeking to overturn Special Relativity and has actually included Nobel Prize Winners in its membership.

Many claim to have theories, tests, data and mathematics that match empirical data that do not require Special Relativity or prove it is false but they are ignored or ridiculed.

However, in all fairness to Einstein modern science also in many cases misquote Einstein and extrapolate Special Relativity into meaning things he did not claim.

For example, it is routinely stated that Einstein proved there is no ether in 1905 yet here is an extract of his speech in Leiden

15 years after publishing Special Relativity and upon releasing General Relativity.

********************** Extract from Speech ********************

Recapitulating, we may say that according to the General Theory of Relativity space is endowed with physical qualities, in this sense **therefore, there exists ether.** According to the General Theory of Relativity **space without ether is unthinkable,** for in such space there not only would be no propagation of light, but also no possibility of existence for standards of space and time (measuring rods and clocks), nor therefore any space-time intervals in the physical sense. **But this ether may not be thought of as endowed with the quality characteristics** of ponderable media, as consisting of parts which may be tracked through time. **The idea of motion may not be applied to it.**

Albert Einstein, Leiden Lecture - 1920

****

It is rather clear that Einstein has not rejected the idea of ether although his ether may not be the original concept of ether. What Einstein said instead was that we have no way of measuring the ether and cannot use it as an absolute frame of reference to establish motion.

Further modern science routinely claims that the Michelson & Morley experiment had a null result proving there was no ether. While the M&M experiment did not support the initial predictions for static ether and was the emergent cause for relativity, the fact is it was not null but contains a diurnal cycle (24-hour cycle concurrent with earth's rotation) consistent with ether of some type.

Several experiments since and recently have produced similar diurnal cycles in their data. This rather suggests a different ether, perhaps a dynamic ether and not absence of ether.

Dan Keith McCoin

Relativists argue the mathematical view of things and not physical view. That is for example they claim Special Relativity is a special case within General Relativity and claim if General Relativity is valid therefore Special Relativity has to be valid as well.

The truth is had General Relativity preceded Special Relativity, Einstein likely would never have conceived Special Relativity. This can be inferred from the following:

*********************** Extract from Speech********************

Quote by Einstein - Chapter 22
General Relativity

"In the second place our results shows that, according to the general theory of relativity, the law of the constancy of the velocity of light in vacuo, which constitutes one of the two fundamental assumptions in the special theory of relativity and to which we have already frequently referred, cannot claim any unlimited validity. A curvature of rays of light can only take place when the velocity of propagation of light varies with position.

Now we might think that as a consequence of this, the special theory of relativity and with it, the whole theory of relativity would be laid to dust. But in reality, this is not the case. We can only conclude that the special theory of relativity cannot claim an unlimited domain of validity; its results hold only as long as we are able to disregard the influences of gravitational fields on the phenomena (e.g. of light)"

Albert Einstein - (A few Inferences from the General Theory of Relativity)

****

What he has just said is that Postulate #2 in Special Relativity (the invariance in light velocity) cannot claim any unlimited validity.   Special Relativity is only valid in absence of gravity and since gravity prevails in every cubic inch of the universe, Special Relativity does not actually exist.

However, he notes that if gravity is sufficiently weak to be able to ignore its affects.   Then Special Relativity may have some utility.   However, being useful and being valid are two distinctly different issues.   It is like saying "If a tree falls in the woods and nobody is there to hear it then it doesn't make a noise.

So at best, Special Relativity may have utility in weak gravity fields but it is NOT valid physical theory.

Another fallacy promoted by many modern scientists is that the Global Positioning System (GPS) proves Special Relativity. It certainly supports the view of General Relativity but does not even use Special Relativity even though numerous educated scientists routinely claim it does.   Frankly, it is either their arrogance or ignorance talking not the reality of the situation.

GPS does compute orbit velocity time dilation but the method used is not Special Relativity but is a form of Lorentz Relativity. Lorentz believed in an absolute rest frame and he developed much of the mathematics used by Einstein.   The length contraction and time dilation formulas used in Special Relativity are in fact Lorentz formulas and the theory relies upon Lorentz Transforms.

The difference in the theories is that Einstein eliminated the absolute rest frame and replaced that with inertial frames. Anybody traveling in an inertial frame can consider himself at rest. Even when you have two frames with inertial relative velocity, they both can declare they are at rest.

Of course you must arbitrarily stipulate one is at rest and the other has motion to compute the relativistic affects but he claimed either has the right to make that declaration. There is no such thing as "Actual Velocity" in Special Relativity. However, in the GPS they do not compute relative velocity between clocks but use a preferred reference frame.

It is preferred because it prohibits the reciprocity affect inherent in the relative velocity view of Special Relativity. The frame used by GPS is called the Earth Center Inertial or ECI frame. Using the ECI frame as a reference it is not possible for an observer in orbit to declare he is at rest such that the center of the earth has orbit velocity.

In that, method the orbiting clock has absolute motion to the center of the earth but the ECI has no relative motion to orbit. This is a technical difference but important to know in any case, because far too many times credit is wrongfully attached to tests claiming they prove Einstein's view when in fact they do not.

Such tests may have results consistent with Special Relativity but if the results are not exclusive to it then it does not prove Special Relativity. If the results have an alternative explanation then the test only supports a defined relationship and not one theory over another.

Claiming ones test results prove Special Relativity seems to be a matter of glory hunting, or "coat tail hangers on" syndrome, to be associated with Einstein's work.

The simple fact is there is an unsettled argument in modern science as to if Special Relativity can or even should be applied in GPS. The reason is orbit is a rotating frame and the fact of rotation means velocity is under constant angular acceleration, which is considered non-inertial, and a frame to be computed by General Relativity.

However, in the case of orbit, it is actually in constant free-fall and free-fall in a gravity field is inertial. Einstein's Equivalence Principle (EEP) states gravity and acceleration are equivalent. The issue comes down to what are called tidal forces in orbit. Gravity pulls toward the center of the earth so any object with a tangential length to orbit is being pulled toward its mid-section or is under a slight compressive force.

Primary GPS orbits are just over 12,500 miles out. If you are in a space craft 30 feet long oriented tangential to orbit the ends of your craft create lines of gravity that form an angle to the primary line of gravity that is 2.6E-5 degrees and generates a compressive force that is 1/2,208,533th as strong as gravity toward the earth itself.

In addition, orbit is not "Zero" gravity but "micro" gravity because net forces are only zero at some finite radial point from earth. Any object with two dimensions in the radial plane must have mass that is both inside the orbit balance point and further out than the orbit balance point.

The orbiting mass stays in orbit because the net pull both inwards and outwards is equal. As such, some mass has a slight centrifugal force while other mass experiences a slight centripetal force and therefore is not in "Zero Gravity" but "Micro-Gravity".

However, as Einstein himself said "If gravity is sufficiently weak so that it may be ignored then Special Relativity may be applied".

Since in GPS tidal and micro-gravity forces are sufficiently weak that they may be ignored then orbit may be considered inertial. But relativists like to avoid that issue, because if you take the view that earth's equatorial surface velocity is 've' and orbit velocity is 'vo' and you set relative velocity between clocks as vr = vo - ve.

Then you try to use vr to compute time dilation using Special Relativity you get a value of 5.8μseconds /day time loss, which is not what empirical data proves. μ means micro or millionths.

If on the other hand you compute time dilation to the common ECI rest frame for 'vo' and then compute time dilation of 've' to the ECI and take the differential between the two time dilation calculations you get 7.2 μs/day time loss.

-7.2 s/day happens to be a correct value. The correct value then is not based on relative velocity between clocks but absolute velocities of clocks to a common rest frame.

However, GPS does not compute the surface velocity time dilation and subtract it. Time dilation of surface clocks due to rotational velocity is only 0.1μs/day. Relative velocity between clocks on the surface and clocks in orbit frames are under constant change because they have different angular rotation.

In addition, it turns out surface clocks at sea level all tick in unison regardless of latitude. Clocks at different latitudes have a different velocity due to the shape of the earth being an oblate spheroid because of the centrifugal affects of rotation. However, the radial distance from the ECI to the surface is greater at the equator than at the poles and General Relativity gravity affect is precisely equal and opposite to the velocity change with latitude and cancels relativistic affects.

The bottom line is some modern day physicists' claim Special Relativity is used and proven by GPS while others say it is not or cannot be used in GPS.

Some claim orbit is inertial others say it is non-inertial.

All will say the ECI is not a preferred frame, that there are no preferred frames, but that is because they use a strict physics definition for preferred. A preferred or privileged frame in physics means a frame where the laws of physics might appear to be measurably different. According to Special Relativity, physics are the same in all frames and hence there can be no preferred frames.

However the ECI is preferred in the standard use of the term meaning "to give preference or priority to" and the ECI gives priority to the orbit by forcing it to have velocity while it's rest frame cannot make the same declaration as it can in a mere relative velocity view in Special Relativity.

The bottom line is that GPS uses the ECI which precludes the reciprocity inherent in Special Relativity's relative velocity view of physics. The orbiting clock cannot be set at rest and have the ECI frame have orbit velocity. It is a technique to stipulate which frame has "Actual Motion" an absolute concept, not merely "Relative Motion" a relative concept. It is a form of Lorentz Relativity not Special Relativity.

Further ALL relativists will argue they are right even though they are in disagreement with each other.

Finally if two events occur simultaneously at a common location, for example two sticks of TNT with identical fuses are laying together so the fuses terminate at the same X, Y location and the fuses are lit at the same time they will detonate simultaneously and they must also detonate at the same time from any frame view.

That does not mean at the same-recorded clock time. There can be a simultaneity shift in when they are seen to detonate. However, when detonation occurs it must still be simultaneous for observers in the X, Y plane because the event has the same X, Y spatial location.

However, due to Special Relativity's velocity addition where 2+2 = 3 mathematics that standard physics rule is violated.

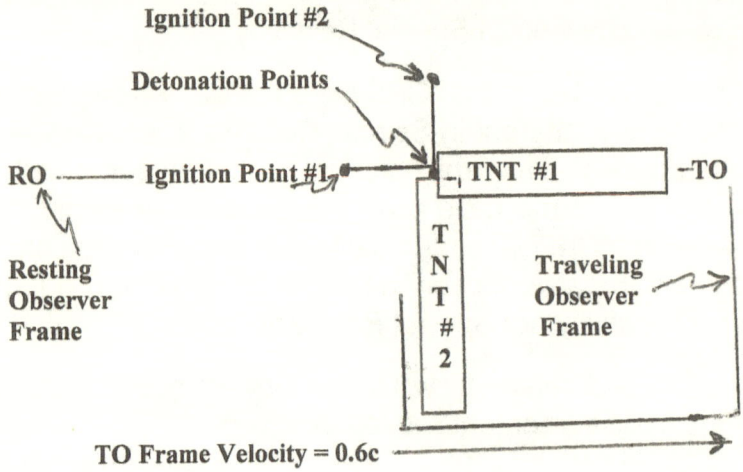

Fig 1b: Prohibited Simultaneity shift caused by Velocity
Addition

The length of the fuses are "L" = 1m. To simplify the problem
alter the speed of light from 3E8m/s to just 3m/s. Set fuse burn
rate (BR) to 1 m/s or c/3.

Two TNT sticks are simultaneously lit by the observer in the
TO frame but for the observer in the RO frame TNT#1 appears
to have been lit t = L/c = 1m/3m/second = 0.3333 seconds
sooner than TNT#2. Due to the speed of light over the length
of the fuse of TNT #1, information received by RO from TNT#2
arrives later, not simultaneously.

The TNT sticks are aligned orthogonal to each other and
in the X, Y plane of RO with the blasting caps at the same
X, Y location. While in the TO frame both TNT sticks are lit
simultaneously and detonate simultaneously 1 second after
ignition because t = L/BR = 1m/1m/s = 1 second, the situation
differs in the RO frame.

In the RO frame the flame of TNT#1's fuse is moving
forward as the TO frame is moving forward hence according
to Special Relativity you must use velocity addition to determine

the compound flame velocity (BR) along fuse of TNT#1 relative to observer RO.

However, you do not use velocity addition for the fuse of TNT#2 because that flame is moving orthogonal to the direction of TO frame motion and flame velocity is not a compounded velocity to RO.

Where v = TO frame velocity = 0.6c = 0.6 * 3m/s = 1.8 m/s , u is the proper fuse burn rate set to 1 m/s , then w is the compound velocity of the flame receding from RO.

TNT#1 fuse burn rate = w - v.   w Is the compound velocity to Special Relativity.

w = v + u / (1 +{ vu / c^2}) = (1.8 m/s + 1 m/s) / (1 + {1.8m/s *1 m/s/[3m/s]^2}) = 2.3333 m/s away from RO. Since the fuse is moving away from RO at 1.8m/s then the net flame fuse burn rate is 0.53333 m/s along the fuse length "L" .

Since the fuse is 1m, long it takes 1.875 seconds to detonate after ignition. In addition, since TNT#1 appeared to have been lit 0.3333 seconds before TNT#2 then TNT#1 detonates in the RO frame 1.875 sec - 0.3333 sec = 1.5417 seconds after being lit or 0.5417 seconds after TNT#2 detonates.

Introducing the false assumption about length contraction, the fuse for TNT#1 is only 0.8m long and should therefore only take 0.8 * 1.875 seconds = 1.5 seconds to completely burn. Less the 0.3333 second advance ignition = 1.1667 seconds and less the 1 second TNT#2 burn time = 0.1667 second delay in the detonation of TNT#1.

Events that occur at the "same physical location simultaneously" must occur simultaneously in all frames by physics but Special Relativity violates that physics principle.

Fig 4a and 4b in Chapter 5 show that the Pythagorean relativistic function is orthogonal to the vector of motion.

Modern physics proclaims that electro-magnetism proves Length Contraction and Special Relativity. I repeat that UniKEF supports relativity but just not Einstein's relativity nor length contraction in accordance with the gamma function. So what is the explanation for electro-magnetism?

1 - In UniKEF, you can have contraction of mass in space. The electromotive and nuclear forces contract according to the Pythagorean function but not contraction of space at that same magnitude but at a much lesser amount perhaps as little as $1E-17^{th}$ if at all.

2 - A mass held stationary in a flowing gravity field develops weight, a measurement made held stationary near a current carrying wire senses a magnetic field.

3 - A free falling mass in a gravity field has no weight; a measurement made co- moving with the current in the wire senses no magnetic field.

This suggests electro-magnetism is simply the result of differential motion in the electric field and has nothing to do with length contraction.

There is no test of the affect on an observer moving at various velocities to the current carrying wire just the rest and co-moving conditions. Those terminal conditions may agree with a view involving space contraction but that fact proves neither spatial contraction nor Special Relativity as the cause.

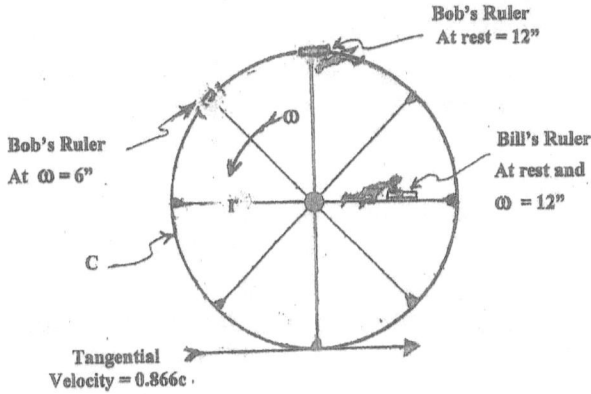

Fig #1c: Relativistic Merry-Go-Round

Another example of Einstein's great capacity to turn otherwise intelligent minds into mush is the "Merry-Go-Round geometry argument illustrated in Fig #1c. He introduced the argument and for over 100 years, no scientist has even considered how ludicrous the argument is.

Brian Greene carries it forward even as late as 1999 in his book about String Theory entitled "the elegant universe". The argument goes that Bob crawls along the outside rim of a Merry-Go-Round with a ruler and for example measures the circumference as being 62.83185307 feet while Bill simultaneously crawls along the radial spoke and measures the radius as 10.00 feet.

They discover that the ratio of C verses r means the circumference of a circle in Euclidean Geometry can be mathematically expressed as a function of Pi, which is symbolized by π.

π = 3.141592654 in a non-ending sequence of decimals places.

ω Is the symbol for rotation radians and a full revolution is 2 Pi radians.

Circumference then becomes C = 2 * Pi * r. However, Einstein and modern science, for over 100 years have argued that when the Merry-Go-Round is rotating such that ω results in a relativistic tangential peripheral velocity for example of v = 0.866c, both time dilates and distance contracts by 50%.

According to Special Relativity, Bob's ruler is now only 6 inches long and he will measure the circumference to be 125.6637061 feet. The radial spoke is not moving in its length vector and does not foreshorten; hence, Bill measures the radius of the Merry-Go-Round as still being 10 feet.

If we now transpose the formula for calculating the circumference of a circle to compute:

$\pi = C / 2 * r$

Computing the circumference we find that π = 125.6637061 feet / 2 * r = 125.6637061 / 20 feet = 6.28318507 or Pi has doubled in numerical value. A completely new set of geometry is claimed because of this conundrum.

What seems unimaginable is that in all these years, and the thousands of intelligent higher educated minds, no one seems to realize that this is simply not true.

It does not take a genius to realize that should I construct the Merry-Go-Round out of 12-inch rulers that the circumference will also foreshorten and Bob will still measure the circumference as being the same as if it were at rest and Pi does not change.

I can use 1-inch rulers or 0.000001-inch (a micron) long ruler to measure both the circumference and radius. The circumference would still contract by 50% and Pi does NOT change. Now I would argue that Bob's ruler may appear shorter to a resting observer standing along side the rotating Merry-Go-Round via Penrose-Terrell Rotation but then so must the circumference appear contracted.

Likewise, the resting observer's ruler may appear foreshortened from Bob's perspective. However, as with other Special Relativity affects the observers perception is not a good measure of physical reality of the other frame. Geometry may appear to change for remote observers' but that has no influence of the physics of the frame being observed.

In the final analysis relativity in both Einstein's view and my UniKEF view have mathematical versus physical reality issue conflicts. In Einstein's view time is dilated in the rest frame but not the moving frame in it space and matter contract. However, I have shown that such space (distance) contraction does not allow for the asserted aging differential.

In UniKEF, clocks physically dilate to account for permanent lost time when clocks are compared in a common rest frame, after having had a period of relative velocity because space contraction fails to provide any differential aging.

However, that causes computed relative velocity to lose symmetry. That is to say while I'm going 100 Mph relative to you, you could be going 50 Mph relative to me.

That also opens the door to FTL travel unless matter contracts as in UniKEF so as to no longer be physical in this universe and seems in conflict with time since it should by relativity have gone to zero at v = c.

Now in the Merry-Go-Round case you have the claim by Einstein that Bob's ruler contracts but the circumference did not making Pi get larger.

There would seem to be no realistic geometry where the circumference could enlarge without the radius also increasing.

Mathematically they treat the problem with + or - curvature but that is an abstract solution not easily translated into physical reality. In the real world there is no cause for a translation of the

X Y-axis radial plane to curve up in the Z-axis to accommodate the mathematical projection.

Try imagining being the resting observer and see Bob's ruler and the Merry-Go-Round circumference contract while the radius remains at a fixed length. Mathematically that can be projected using the 3rd dimension but the Merry-Go-Round is physically constrained to only two dimensions in the radial plane.

The real conflict however, becomes one of physics. Physics have been declared to be the same in every frame and in Bill's frame when the Merry-Go-Round is rotating he will experience an increasing centrifugal force as he crawls outward along the radial spoke

However, in non-Euclidean Geometry, from the rest observer's frame of reference the circumference contracts causing the radial spoke to curve upward and back toward the axis of rotation. You can use the fixed radial length and compute centrifugal force as not changing however, by geometry the end of the radial spoke at the circumference is now less distance from the rotating axis.

In that view centrifugal force would begin to increase, stabilize and then begin to decrease such that at a peripheral velocity of $v = c$ centrifugal force at the rim would have to be zero because there is no circumference and hence the radial spoke end is back at the rotating axis.

Yet without at least matter, contraction in UniKEF the Qualitative Domain Limit does not exist. This situation mandates a new thought approach to the conflict between our mathematics and physics. It would appear not only logical but necessary that there IS an absolute universal rest frame and that some affects we see are caused by motion relative to that frame but that time dilation may merely be clock dilation or the changing of the marking frequency along a universal absolute time line.

Another fallacy promoted by Einstein is called the EEP or Einstein Equivalence Principle. In it he claims that acceleration and being in a gravity field are identical and should be treated as such. That one cannot tell if he were in an accelerating elevator or standing in a gravity field.

That is blatantly false physics. Gravity has an inverse square force relationship to distance where acceleration is constant over distance. Placing simple load cells or spring scales vertically along the wall of the elevator tells you immediately that the force is constant (acceleration) or decreasing with height (gravity).

He would have you believe that gravity is as though the surface of the earth is accelerating upward under your feet pushing up against you. He calls centrifugal force a fictitious force. Ever get thrown off a Merry-Go-Round by this fictitious force? Seemed real enough to me.

But as with other fictitious basis for his theories Einstein wants you to ignore the reality and pretend that you are just an infinitesimal spot with no distance over which to actually detect this difference - NOW THEY ARE THE SAME. But considering them the same out of context of reality is simply nuts.

If you continue to reduce time and distance of an acceleration period you have less and less change in velocity. At zero time you have zero change in velocity. The argument is that with no change in velocity there is no force because $F = ma$. With no Δv there is no 'a' hence no F. That too is nuts.

The force of acceleration is still there but just no motion. It is the static force that produces the acceleration in the real world, unlike the mathematical playground they enjoy. It is the same as weight in a gravity field. The mass isn't accelerating along with the moving gravity field and that produces the force of weight.

Gravity and acceleration are not equivalent and should never be considered and treated so. They extrapolate this distorted reality to mean acceleration has variable affects on time over distance. Another case of bending reality to fit theory.

Finally, the Granddaddy of claims by relativists is that cosmic muon time dilation proves Einstein's relative velocity view. The claim is because according to their statistical life span and velocity they should only travel 660m but actually make it to earth's surface due to relative velocity time dilation.

However, the following tests have been done to compute the earth's "absolute" cosmic velocity; including one study based on cosmic muon anisotropy. Anisotropy is the difference in muon flux from other directions.

1 - Galactic Red Shift: Ground, Balloon and Airplane tests

de Vaucouleurs & Peters, 1968;		300m/s ± 50 m/s
Couklin	1969;	200m/s ± 100m/s
Henry	1971;	320m/s ± 80m/s
Rubin et al	1976;	600m/s ± 100m/s
Smoot et al	1977;	390m/s ± 60m/s

2 - One-Way Light Velocity: Mirrors and Slotted Disk Tests

Marinov	1974;	300m/s ± 20m/s
Marinov	1984;	360m/s ± 40m/s

3 - Muon Flux: Measured Anisotropy of Muon Flux at the Earth's Surface

Monstein & Wesley	1995;	359m/s ± 180

Notice that the muon test using slight muon anisotropy virtually matches Marinov's latest test except for the large ± factor. Taking the ± 180m/s into account their results predict a velocity of 179m/s to 539m/s. Certainly a wide range but well within the range of all other test methods predicting the same absolute type motion.

All these tests resulted in the same general direction of calculated absolute motion. Eight tests using very different methods that all produce the same direction of motion are ignored because they fly in the face of the Special Relativity proclamation that there is no absolute rest frame. A proclamation based on absence of evidence not evidence of absence.

These tests require relativity time dilation to be based on an absolute velocity of the Earth through space and hence relativists ridicule the tests. They point to the ± range of the results and claim they prove nothing.

Granted they are not conclusive but clearly, they merit proper consideration and not just be blown off because they challenge Einstein's relative velocity view. However, the fact is out of thousands of physicists few have picked up the challenge and replicated the test to try to improve on the data.

The simple fact is muon flux does support the time dilation argument of Special Relativity. However, not exclusively, so it does not prove Einstein but rather suggests relativistic affects are based on an absolute motion, which would falsify Special Relativity as the cause.

Such follow up tests have not and will likely not be done because there is no interest in the truth since Special Relativity has utility as it is. That is not science or physics; it is a biased cop out.

You should keep in mind that the assertion all this proves Einstein is also based on some loose statistics. First, the decay of particles is a statistic. Second most muons are generated at altitudes over 9.6 miles up and require a relativistic gamma factor equal to 24, which corresponds with a velocity of 0.999c.

Earth's atmosphere extends up to approximately 75 miles and for muons generated there it would require a relativistic gamma of 187 or a corresponding velocity of 0.9999857c.

Therefore, the muons arriving at earth's surface based on Special Relativity have a relativistic gamma ratio of about 8/1. A much wider deviation than the 359m/s / 180 m/s = 2/1 of the calculation of absolute motion made by Monstein & Wesley based on muon anisotropy.

So a velocity change of only 0.000985c (1/1,015thc) varies the distance traveled from 10 miles to 75 miles! Yet this data proves Einstein? While anisotropy of muon flux calculations showing an absolute velocity in the correct cosmic direction to within reasonable ranges of (7) other test methods, is proof of nothing?

They can claim there is no bias but talk is cheap and facts are rather clear on that issue.

A number of things are completely disregarded when claiming muon flux proves Einstein's view.

1 - Cosmic muons are arriving at earth at near light speed from every steradian direction. It is obvious therefore, that it is not the earth moving at light speed to the muons, which would require the earth to move in every direction simultaneously.

The 3D muon flux mandates that the muons have actual motion to earth and not that the earth is moving to all the muons. So the Special Relativity argument that the muons can claim to be at rest and that it is the earth that is moving is falsified.

All of Special Relativity requires you ignore the physical reality around you and assume special conditions such as being the only two frames in the universe that exist for purpose of your calculation. i.e. Having you consider only one muon and the earth, disregarding other muons arriving concurrently from other directions.

2 - There is no aberration or parallax of muon flux arriving orthogonal to the vector of approach between the earth and muon in question.

3 - It was the anisotropy of the muon flux in the above1995 test that mathematically yields the calculation of earth's actual

absolute motion in the universe. Once such tests can fix the numerical value more precisely, and repeatedly, we will then have an absolute rest reference, which is prohibited in Einstein's Special Relativity.

4 - There is no test or data to support the claim in Special Relativity that from the muon's perspective the earth clocks dilate and lose time. That is there is no support for the impossible claim in Special Relativity that inertial velocity is physically in fact equivalent to being at rest. Not being able to sense or measure motion does not mean you are in fact at rest. In other words, there is no proof that absolute rest does not exist.

5 - When considering relativistic affects the simple fact is, that at everyday velocities we are substantially at absolute rest. Even at the solar systems 360m/s motion in the cosmos is still 360m / c = 0.0000012 or 1/833,333th of c.

At that velocity, the relativistic gamma would only cause a time loss of only 6.2E-8 seconds per day or it takes 44,041 years to lose 1 second. Do not forget leap year in your calculation that yields a 1 second loss every 44,012 years!

For relativity purposes, we are virtually at absolute rest. Only by more testing specifically looking for a shift in relativistic velocity data where for example at 0.99999c one gets γ =223.607356769 in one direction but gets γ = 223.607356489 in the opposite direction will we exclude or confirm an absolute rest condition. This procedure is to better identify our absolute motion in the universe and affix an absolute rest reference.

Special Relativity begins to make some sense if you assume we are at absolute rest and that motion in any direction is an absolute energy level change universally. Of course, we are not at absolute rest and that confuses the issue even though the velocity we may have seems to be so small as to have immeasurable mathematical affect relativitisticly.

CHAPTER 4

UniKEF Introduction:

UniKEF is an informal concept of alternative explanations for gravity and relativity. It is not intended to claim any definite validity but is presented as one consideration to resolve modern physics conundrums.

UniKEF stands for "Universal Kinetic Energy Field". It assumes space is created by a flowing energy. The energy must effuse from every spatial ordinate point and flow in every direction.

It is the most basic thing in the universe. To date it has been directly undetectable but is inferred by its affects. Those affects are physical changes that require flow of energy, producing our concept of time and production of gravity.

It is unique and very complex. When it comes into existence, it pushes the energy that just preceded it as demonstrated by Fig #2. 1 becomes 2 and 2 spatial points then push out energy becoming 4, 8, 16, 32, 64, 128, etc.

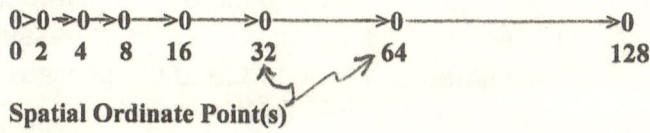

Spatial Ordinate Point(s)

Wave Velocity versus Increments of Energy Released

The wave front rapidly accelerates to and beyond v = c. While nothing physical may become infinite it reaches toward infinity forever going faster and faster.

This wave accounts for particle entanglement called "The spooky action at a distance" by Einstein. An affect in physics where things happen faster than the speed of light (FTL).

An entangled particle light years away can change its spin immediately when its entangled pair partner changes its spin, faster than the Special Relativity v = c limit allows.

Modern science salvaged Special Relativity, which prohibits FTL motion by claiming that no information is transferred in the process. Frankly that seems to be a straw man excuse since they do not know how the process works, hence are only assuming changing spin requires no information.

Claiming no information is transferred is an assumption, which on its surface would seem refuted by the shear fact that the remote particle received notice of the spin change and that is information.

UniKEF does not accept the concept of a Big Bang. The Big Bang claims that all energy and mass in the universe sprang spontaneously from a singularity. A singularity is a point source without dimension. That is the entire universe once existed in a dot so small it had no volume.

Politely put the Big Bang is a ludicrous concept.

That event is then followed by "The Inflationary Period", a time when it is claimed that the universe expanded at superluminal velocity or FTL, so as to be in the conditions we assume had to exist at certain times as we understand the process using a Big Bang inception and looking back in time.

UniKEF rather has the universe come into existence by what I have labeled as being a Big Rip. That is a tear or Brane surface eruption, which produced a larger volume immediately requiring no inflationary period and which precludes the concept of a singularity condition of infinite mass and energy density in zero volume. Perhaps a bounce back after a prior contraction in an oscillating universal history.

While I clearly cannot describe the how or why it all happened, I do believe the following mathematically describes the process. The universe came into existence via an ex-nihilo process where (N) is Nothingness and (S) is something. (S) Comes in two varieties + and - .

(N)------------------> (+S) + (-S). It can be seen if you assign any opposite numerical values to (S) then you can have 0---------------->(+1) + (-1). We exist by a bifurcated nothingness process. This avoids the concept of creation since nothing has been created. On balance, the universe equals a zero existence. We exist by borrowed energy.

This is not an easy concept to grasp but there are a couple of known events in modern science that suggest it at least seems plausible and it does produce many results that provide suitable alternatives to the current relativity failure.

We currently see virtual particle pairs +/- in nature come into existence and in a short time span vanish via an ex-nihilo process. UnRuh demonstrated that under high acceleration virtual particles become real particles by taking energy from the accelerating observer.

Stephen Hawkings mathematically showed that virtual particles near a Black Hole could end up with one becoming captured by the Black Hole leaving the other to drift away and become a real particle. The process is called Hawkings Radiation.

UnRuh particles and Hawkings Radiation come from nothing and become something. So as a minimum the concept of ex-nihilo existence is confirmed. What has been missed I suspect is the fact that the lifetime of virtual particles to not violate conservation of energy is linked to their mass.

If you consider the mass of the universe and correlate that to a lifetime not violating conservation of energy, you get many billions of trillions years left in the universe's existence.

In fact, it generally correlates to the decay life of a proton, which probably is not a coincidence. The half-life of a proton has been calculated to be between 1E31 to 6.6E35 years. That is 660, 000, 000, 000,000,000,000,000,000,000,000 years! Latest estimates are the universe is currently 13,700,000,000 years old.

That is 48,175,182,248,000,000,000,000,000 times as long as we have existed to date. That is just the " Half-Life" so the universe should be around for quite some time.

However, unless humankind develops interstellar space travel and finds a suitable new star and planet system, we cannot survive more than another 5 billion years here on Earth in any case. We are at 50% consumption of our suns nuclear fuel as a yellow star.

It will then become a Red Giant that will expand in size approaching to about 95% of earth's orbit. Earth will be consumed by over 5,000 F heat of our Sun's photosphere.

Space is flowing energy and mass is compacted space. Space therefore has a fabric density approximately equal to mass times 1.1E-17. The problem is in knowing what pure mass density is. For example, a lead ball is virtually all space with very little mass.

Its weight is within the sub-atomic particles and those are made up of quarks. If string theory has, any merit quarks are made of strings. Therefore, the space volume to mass volume ratio is not readily apparent.

If this " . " were an electron and this " O " is the nucleus of an atom, they would be located 210 feet apart. Another analogy is if the nucleus of an atom were 150 feet in diameter (half a football field) and located in Los Angles, the surrounding electrons (Bohr Atom concept) would be located in New York City.

What make things seem solid are the electromotive forces between electrons. Electrons in your finger repel electrons in matter and it feels solid but is actually substantially nothing but space.

CHAPTER 5

Consequences of UniKEF

The flow of UniKEF energy creates the bounds of our universe. Recalling that the UniKEF flow is effusing from every spatial ordinate point, every spot in the universe has energy flow of v = 0 to virtually infinite velocity.

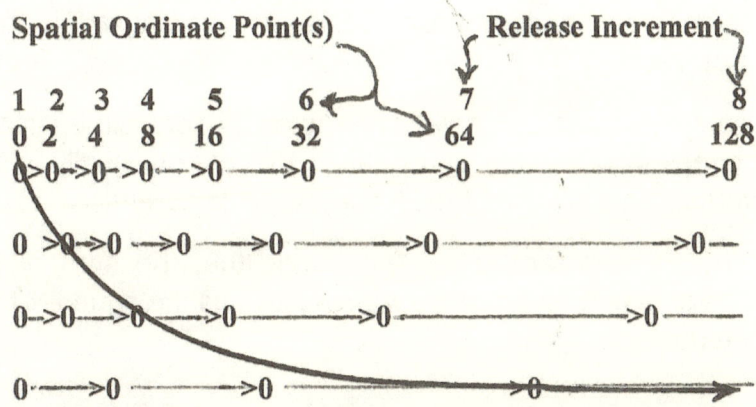

UniKEF Energy Levels at Spatial Ordinate Increments

It can be seen that position #64 has UniKEF energy flowing at values equal to the origin energy of all other ordinate positions surrounding it that are virtually infinite in number and hence possess a full range of energy including virtual infinite velocity.

In UniKEF, the Pythagorean Theorem comes into play in that in your inertial rest condition you are at a specific universal energy level. Dimension, time and distance are functions of the amount of UniKEF at your inertial frame energy level that surrounds you. It is called the "Quantitative Energy Domain".

UniKEF has a minor density and interacts within itself. That is as it flows through other UniKEF there exists an interference or resistance and energy becomes dissipated creating a diminishing amount of energy in your Quantitative Domain over distance.

This creates a finite boundary to your universe putting you at the center of the observable universe. This was a priori prediction, which is now confirmed by red shift analysis of the universe.

Lorentz concluded that dimension of an object contracts in the direction of motion. The underlying principle is that our universe is constituted by UniKEF energy flowing within the v = c boundary.

At v = c mass will have contracted to zero dimension in the direction of motion and ceases to exist physically in our universe. This involves the Pythagorean Theorem.

The Pythagorean Theorem states that the sum of the square of legs o and n of the triangle equals the square of the hypotenuse m.

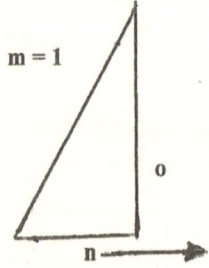

Pythagorean Theorem

Fig 4a

Where 'm' is time or dimension, velocity is 'n' in terms of v^2/c^2 and 'o' is the relativistic affect. 'v' is velocity and equivalent to a universal energy level called "Qualitative Domain" in UniKEF. The Qualitative Domain Limit is physically at $v = c$.

Mathematically:

$$o = \sqrt[\Lambda^2]{m^2 - n^2} \quad \text{and since } m = 1: \quad o = \sqrt[\Lambda^2]{1 - n^2} = \sqrt[\Lambda^2]{1 - v^2/c^2} \ .$$

This is the basic relativistic term $1/\gamma$ (1/Gamma). In Special Relativity the example is given where a light beam is bounced between mirrors at the ends of side 'o'.

The observer traveling 'v' in the vector 'n' will measure $v = c$ for the light beam but an observer at rest on side 'n' sees the light beam travel along path 'm' which is greater distance but takes the same time.

The consequence is the relativistic term Gamma for time or distance differentials between frames.

In Galilean Relativity light traveling along 'm has a velocity of $v' = c + v$.   That is the beam has the standard $v = c$ of light orthogonal to frame vector of motion but also contains the frame velocity component v along vector 'n' so the compound velocity to the resting observer is considered and Gamma does not exist.

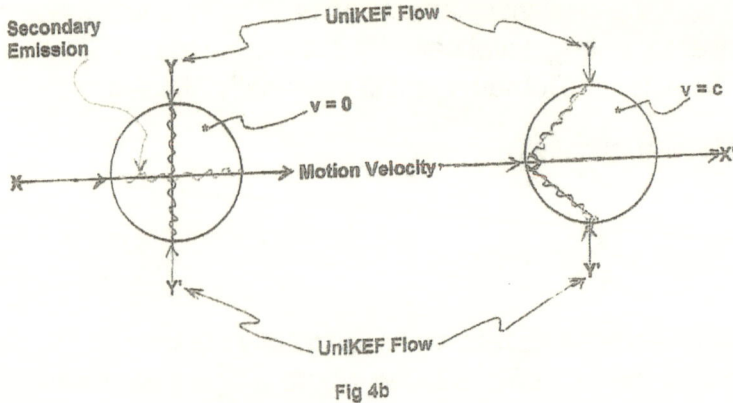

Fig 4b

Motion through the UniKEF fabric of space produces a parallax of energy flow from the Y vector versus the X motion vectors. The Y vectors reproduce the Pythagorean relationship as velocity varies from zero to v = c. It is as though time and dimension of mass are related to an orthogonal secondary emission from a collision of the primary UniKEF fields.

This is consistent with the idea that relativistic affects are the consequence of shifts in absolute energy levels between objects and a common spatial fabric energy reference and not just between two objects.

In UniKEF mass does contract according to current Lorentz formulas but space does not.   Space may contract but at a substantially lesser rate. Just as space is energy and mass is compacted space the E = mc² ratio suggests that space may only contract at a rate of 1.1E-17th as fast as mass.

Atoms are held together by a variety of atomic and electromotive forces that are non-existent in space or are substantially stronger. Contraction of matter is a function of those forces and not of contraction of the internal space. Matter contracts in space and is not because its internal space contracted.

The Quantitative and Qualitative Domains suggests that there may be multiple universes, much like bubbles in a boiling pot of water. They may exist in different sizes, different energy levels and different ages; even perhaps might occasionally be merging.

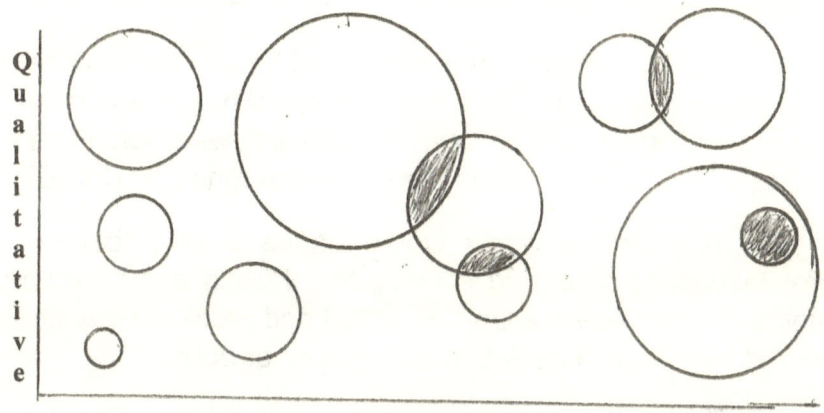

Quantitative

Multiple Universes

Fig #5

PRIORI PREDICTIONS Based on UniKEF

1 - Multiple Universes: Such potential mergers lead to a priori prediction that we might find remote areas in our universe that seem older than our universe at large.

STATUS: Such objects have been observed but have not yet been linked to merging multiple universes although many are now considering multiple universes.

2 - Massive gravitating bodies would be heated internally by the production of gravity.

STATUS: NASA has observed this phenomenon.

------------Science Today, April 1964 issue, GEOLOGY --------

"For example, a global analysis of heat flowing to the surface from the planets interior shows remarkable similarity to a world wide analysis of the earth's gravitational field from satellite observations. Scientists have no idea why the patterns should be similar and have gone to work on the problem."

A thin mantel would conduct internal heat to the surface in an inverse proportion. That is less material hence less gravity would result in more heat flow. More heat flows where there is stronger gravity and that fits the prediction made by UniKEF.

3 - A gravity shadow would cause a perturbation in gravitational force during an eclipse. I used a rudimentary method to calculate the UniKEF affect and predicted that there should be a 4.2E-9 deviation during a lunar eclipse.

STATUS: the Geodetic Institute in Frankfort, Germany, discovered the year I began UniKEF a deviation of 4.28E-9 during a lunar eclipse in Norway. I learned of the discovery 13 years later - See Appendix "A".

I received permission to include their findings in my manuscript in 1971, 17 years after I made the prediction. See Appendix "B". The accuracy of the shadow prediction however was pure luck in that the process I used was a very general representation of the UniKEF view.

4 - I predicted that the expansion of the universe was accelerating.

STATUS: That is now widely accepted as fact.

5 - I predicted we would find that we appear to be at the center of the universe.

STATUS: Red shift analysis of the universe makes it appear to be the case.

************************* Paper Extract *************************

IS THE EARTH THE CENTER OF THE UNIVERSE?

Abstract: Varshnim Y.P.: 1976, Astrophys.Space Sci., 43, 3.

It is shown that the cosmological interpretation of the red shift in the spectra of quasars leads to yet another paradoxical result:

Namely, that the earth is the center of the Universe.

****

6 - I predicted that the $v = c$ limit imposed by Special Relativity would be violated and we would find objects moving Faster Than Light (FTL).

STATUS: This has been observed but has been countered by relativists. Hundreds of objects have been found to appear to be moving FTL. However, when FTL objects were discovered and threatened the Theory of Special Relativity, relativists worked overtime to find a solution.

The solution was that mathematically they could show that an object moving near the speed of light along the line of sight would appear to be moving FTL even though it was not. The affect was caused by an "Illusion of Motion". However, this only happens in a narrow angular range of motion along the line of sight.

The problem is that motion along the line of sight also produces either a red or a blue Doppler shift in the frequency of light and less than 1% of observed FTL moving objects have red or blue shifted data. That means they are not moving in the line of sight area where the illusion applies. However, the inapplicability of the illusion is ignored and observed FTL motion is not explored further or considered as a challenge to relativity.

7 - UniKEF was developed on the idea that space is not just a vacuum void but is created by and filled with energy.

STATUS: It is now generally accepted that space is packed with intense energy.

8 - UniKEF predicts that there is a limit to gravity as mass is continually added hence Black Holes do not form singularities. There was no Big Bang followed by an inflationary period where expansion exceeded the speed of light but more of a Big Rip over some plane similar perhaps to today's view of a Brane in string theory creating an instant volume.

A singularity is where all the mass and energy of the universe exploded from a place that had "Zero Volume". A singularity is a hypothetical mathematical extrapolation, which ignores the probability that there are physical barriers to reaching such ludicrous conditions.

It is the same principle as measuring the rising temperature of water and predicting that it will reach 1,000 F without being pressurized because you do not know about the vaporization phase of water where it becomes steam.

The UniKEF prediction that there is a limit to gravity as mass is continually added is based on flowing energy density of space, which produces gravity in the mass, by attenuation. Once UniKEF flow is totally blocked by mass, continued addition of mass has no affect on gravity potential and therefore Black Holes do not form singularities.

STATUS: Scientific American October 2009 issue, page 38 has an article where researchers say they now believe there is a physics process that precludes Black Hole singularities.

9 - UniKEF predicts that the Universe came into being via an ex-nihilo process Where (N) is Nothingness and (S) is something:

N--------------------> (+S) + (-S)

Putting arbitrary numbers value for "S" we find N-------------> (+1) + (-1) = 0. That is we exist via borrowed energy. Nothing has been created because on balance its value is "Zero". With no creation, there is no Creator. I understand this can be hard to grasp but there is strong evidence it is true.

STATUS: The ex-nihilo process generates +/- virtual particle pairs. Stephen Hawkings has shown that virtual particles can become real particles via Hawkings Radiation; a process where a Black Hole captures one particle and leaves it opposite mate to become a real particle in space.

UnRuh showed that virtual particles become real particles to an accelerating observer. The energy required in UnRuh comes from the accelerating observer.

10 - UniKEF predicts that our universe is finite and is bounded by both a "Quantitative Domain Limit" which sets the rest dimensions and a "Qualitative Domain Limit" which restricts relative velocity (energy) bands of existence in our universe to $v = c$.

In this view, you are in the center of your own universe (See Priori #5). We each have our own universe. If you are 100 miles east of me then your universe extends 100 miles further east than mine does and mine extends 100 miles further west than yours does. This only applies if we are of the same mass (See #12 below) and have the same inertial velocity in space.

It is not a chicken and egg situation. Many ask, "Well if it is finite what is on the other side of the edge"? You have to realize that the edge is created by attenuation of spatial energy (Quantitative Domain Limit) in your 'c' band of existence (Qualitative Domain Limit) and if there is no energy, there is no space and no time. Without space or time, Nothingness is not a void.

Nevertheless, what lies beyond your edge may in fact be more of the same for untold distances. The edge is a physics limit, a point where beyond which things have no dimension and hence no longer interact with your physics. However, as you move towards the edge it moves forward with you. You are not approaching or reaching the edge.

The edge of the observable universe and physical universe are not the same.

The Qualitative Domain Limit is similar but is not based on distance and energy attenuation but is based on a differential energy limit of v = c; such as the Lorentz Contraction of rods with relative motion. At v = c relative velocity objects dimension in the vector of motion to you contract to "Zero" and cease to exist to you physically. This affect however is frame dependant and in the moving frame, all measurements are still the same because rulers have also contracted by the same proportion.

Special Relativity precludes objects from reaching this domain but UniKEF does not.

STATUS: Pending confirmation

11- UniKEF sees the issue of "Relativistic Mass" as being a misinterpretation of empirical data. While it is true that our experience with particle accelerators confirm that it takes more and more energy to cause an object to continue to accelerate, there are two issues that need to be considered before suggesting mass is increasing and becomes infinite at v = c.

1 - Particle accelerators use EM waves to accelerate the particle. EM waves purportedly have a limit of v = c and it should be no surprise that something limited to v = c could not push something else faster than it can go itself.

2 - Just as EM used in a coil puts a magnetic field out into space and restrains current flow initially but produces current flow by collapsing back from space into the coil once the motive voltage force is removed, the EM pushing a particle may lose energy transfer efficiency as it approaches v =c by becoming stored into space.

In other words, what we see as increased energy required to accelerate a particle is in fact a decrease in energy transfer efficiency and not an increase in mass. Further what measures as added momentum in stopping the particle, and appears as being increased mass, is nothing more than the magnetic field stored in space chasing along behind the particle catching up when the driving force is removed, causing an effective flywheel affect of increased push on the particle and is not really from any mass increase.

In such case a rocket where the fuel, thrust engine and load all move together, there is no relative velocity hence no change in respective masses and therefore no limit on universal velocity but only the Qualitative Domain Limit to existence in this universe. However, the rocket would continue to exist normally in a higher plane energy universe, just as at sub-luminal velocity the moving frame sees no relativistic changes in time.

It is unrealistic to believe that other observers' perception has any role in your physics. It only limits the remote observers' perception, not your action.

STATUS: Pending Confirmation

12 - Distance is not a fixed value in UniKEF but varies as a function of observer mass. This is not readily noticeable for small masses or short distances but in cosmic terms, it becomes a major issue. It is comparable to the curvature of space in General Relativity but is based on different principles. The consequences, for example, it is 4.3 light years distance from Earth to Alpha Centauri by light measure BUT for a bowling ball, it is less.

This is explained in Chapter 6 "UniKEF Gravity".

STATUS: Pending Confirmation but parallels General Relativity with an inverse cause. That is General Relativity claims curved space causes gravity and UniKEF claims that gravity causes curved space.

13 - Time is not a dimension in UniKEF it is an artifact of an energetic 3D universe. Past, Present and Future are the brains invention to catalog change caused by energy flow. We live in a DYNAMIC PRESENT in STATIC TIME or an energy dependant universe.

Your dynamic present is different from mine. If you are located on Jupiter's moon Io and I on earth with Jupiter and Io in line with earth. And if another person were located midway between the sun and Jupiter and in his view the Shoemaker-Levy comet impact on Jupiter and a solar flare on the sun occurred simultaneously then:

I will see the solar flare occur in 8.3333 minutes after it actually happened. It and millions of other events information arriving at that instant form my dynamic present. I will have another dynamic present 39.34 minutes after I experience the solar flare that now includes the comet impact on Jupiter.

I use the time term minutes but in reality it is some quanta of energy flow, not time flow per se. No, man made clock measures time rather it merely marks events caused by energy exchange sequentially by some man made interval of increments.

You on the nearby moon IO have a dynamic present 1.4 seconds after the comet impact that includes it and 43.25 minutes later that include the solar flare.

Our entire existence or sense of PRESENT is actually comprised of all PAST and FUTURE events in the universe compared to other spatial ordinate PRESENTS. Such that to move in any direction one must simultaneously enter the PAST and FUTURE from the initial ordinate position.

It does not matter how the event information is delivered. Light propagates over 186,000 miles per second and sound around 1,100 feet per second. Such that to the carpenter his hammer makes noise virtually the instant his hammer hits the nail but if you are watching from 1 mile away you will have a DYNAMIC PRESENT where you see the hammer hitting the nail 5.4 micro-seconds (millionths of a second) after it was actually hit.

You then hear the hammer hit the nail in another DYNAMIC PRESENT while you see the hammer being raised over the carpenters head 4.8 seconds later. Your sense of Present has shifted information from the carpenters DYNAMIC PRESENT. His DYNAMIC PRESENT is in your future and your DYNAMIC PRESENT of these events is in his PAST and vice-versa.

Time is an attribute of energy exchange in an energetic 3D universe. However, energy exchange requires what we mentally perceive as time. Time is not a dimension; it is an attribute of an energetic 3D existence. Without change, there are no events hence no measurable passage of time. Time cannot be traveled backwards; to reverse time you would have to reverse entropy in the entire universe simultaneously.

If you could some how achieve reversing universal entropy, you would merely grow younger until your birth and cease to exist in any prior time. Going into the future involves hibernation or suspended animation, not a modification of time per se. You can never return to the point in time you left without again reversing entropy throughout the universe.

The arrow of time always moves forward.

Aging more slowly due to relative velocity according to Special Relativity's time dilation predictions is highly questionable since no known clock is actually measuring time but is merely marking universal time at different frequencies when universal absolute velocity changes.

We have no empirical data that the physical affect on particle decay or atomic clock frequency are true indicators of living cell division or mortality while in relative motion.

The likelihood is you would arrive at your destination in less recorded time but have not aged any differently than a resting twin observer. See Bob's trip in Chapter 1.

STATUS: Fact known by all but not made obvious.

14 - UniKEF predicts that one day it will be found that the apparent invariance of light is actually an artifact of its quantum energy production and not of its propagation. That is to say, observers moving with different velocity to a light source that measure light speed as being constant are measuring different photons and not a photon with magical qualities to be at different places at the same time or at a place at different times based on observer perception.

There are things that suggest this is true.

1 - Cerenkov Radiation is a process where charged particles traveling through water, faster than light (FTL) for that medium, produces photons, which is the infamous blue-white glow of nuclear power.

Now it is noted that light travels 6.9E7 m/s slower in water, so the particles are not violating the v=c limit imposed by Special Relativity for velocity in a vacuum.

However, atoms are not filled with water, electrons routinely jump from one orbit to another instantly without existing in between orbits, and the result is the production of photons. That is the operating principle of lasers.

The apparent FTL transition of electrons between orbits "inside" the atom produces photons even though atoms are filled by vacuum of space.

However, if you consider space as a medium then anything exceeding v = c or 3E8m/s could produce light, as we know it. Just as particles, traveling FTL in another medium produces light.

2 - This provides a potential solution for particle entanglement as well, since signals or a vibration heterodyned onto the primary UniKEF flow rapidly reaches virtual infinite velocity as in Fig #3.

Therefore the UniKEF carrier medium has information which is traveling past each observer at velocities <>c for the spatial medium but if light production is a quantum energy function then the observers velocity component relative to the source inertial velocity to the spatial fabric varies when and where a photon is produced at v = c and is then observed.

3 - One can think of light as being a form of dimensional binding energy release for objects reaching and exceeding the Lorentz Contraction point of zero dimensions. Effectively Star Trek and "Warp 5 Scotty" then a" *Flash of Light"*.

STATUS: Affects observed but pending Confirmation or recognition as the cause.

15 - UniKEF predicts that the Universe is finite, not infinite in its dimension.

STATUS: This is required for UniKEF gravity to function.

16 - Finally the prediction that Special Relativity would be found to be false.

I do believe my analysis regarding the assertion by Special Relativity in the twin paradox issue that the traveling twin goes 1/2 the distance and accumulates 1/2 the amount of time of the resting twin, hence he arrives back younger proves Special Relativity is based on flawed physics principles and extrapolates mere mathematics to impossible and ludicrous consequences.

Where TT is the Traveling Twin, RT is the Resting Twin and velocity of the traveling twin is 0.866c such that Gamma = 2.000 where d is distance = 1 and t is time = 1 then:

....._____TT____ _RT_

v = 0.5d / 0.5t = d / t = 1.000

It can be seen that if the distance is 60 miles at rest and velocity is 60 Mph but is hypothetically given the function of being 0.866c where Gamma = 2.0, time and distance will be reduced to one half normal, according to Special Relativity then:

According to TT, the trip requires one hour to complete. That is 3,600 clock ticks at 1 tick/second. 3,600 ticks / 60 Miles = 60 ticks / Mile. This is also what Special Relativity asserts occurs in the Resting Twin frame view.

Special Relativity then claims because Gamma = 2.000 that when the Traveling Twin returns his clock will have only accumulated 30 minutes. Further, since he was traveling 60 Mph he must have gone only 30 miles.

However, on detailed analysis you can see that if you accumulate 30 minutes, which is 1,800 seconds in 30 miles your clock ticked:

Tick Rate = 1,800 ticks / 30 Miles = 60 Ticks / Mile. The same tick rate as RT's clock.

Since both clocks have the same tick rate they must tick in unison and remain synchronized such that when TT returns home in 30 minutes RT's clock must also display 30 minutes not the hour claimed by Special Relativity.

STATUS: Physically valid- Falsification pending confirmation.

Based on the numbers of these priori's that have become observed or accepted today and by the falsification of Special Relativity, there is great belief in the general principles of UniKEF even though it is not formally constructed , has some counter intuitive features and without question has components that are invalid speculation.

Special Relativity can still be used but we should continue to seek a better physical theory while remembering that Special Relativity is just a mathematical tool with limited physical validity.

CHAPTER 6

UniKEF Gravity

Newtonian Gravity is expressed as $F = G * m1 * m2 / r^2$. F is the force of gravity, m1 and m2 are the two gravitating masses, G is the universal gravitational constant and r is the distance between the centers of mass.

While mathematically it coincides with a wide range of observations, it suggests "Action without a Cause" and ultimately fails to describe gravity over vast distances.

UniKEF Gravity is a physical model where the cause can be shown and mathematically it provides the function over the entire spectrum of observation. It begins with a local inverse square relationship between F and r. Force decreases with distance of separation.

It then becomes more linear which corresponds with the flattening of the gravity curve at galactic distance scales, that now require the creation of a hypothetical Dark Matter or MOND (Modified Newtonian Dynamics) a pure mathematical treatment to fudge the forces calculated to agree with observation.

Stars in general have been found to be orbiting galaxies twice as fast as Newtonian Gravity can account for. Stars should be flying off into deep space, not circling the galaxy. Observed galactic gravity is much stronger than predicted by Newton using the standard formula.

There is either 4 - 5 times as much mass as we see or the gravity curve flattens and deviates from the pure Newtonian principle. It decreases more slowly over vast distances than $1/r^2$.

Finally, UniKEF mathematically goes through zero and becomes repulsive at cosmic distances which is consistent with an accelerating expansion of the universe and that suggests that the current hypothetical Dark Energy may actually be what I labeled UniKEF back in 1954 but that it also causes gravity not just expansion of the universe.

Dark Matter and Dark Energy have opposite physical affects. They require a unique distribution to produce the observed functions if they exist as currently proposed.

UniKEF unifies the phenomena of gravity into one coherent physical process in lieu of three separate regimes of gravity caused by three different mathematical and hypothetical causes or without cause as the case may be. UniKEF may apply at a quantum level but that has not been considered beyond possibility that the strong nuclear force may be UniKEF enhanced gravity. See Chapter #9.

UniKEF produces gravity by the physical product of energy transferred to cumulative mass by a simple geometric and trigonometric relationship. Then flattens to be stronger at galactic distance scales and finally becomes repulsive at cosmic distance scales.

Where Newtonian mathematics trends toward "Zero" at infinity, the UniKEF inverse square trends toward "Unity".

Rudimentary Process Description

UniKEF is based on the concept of a laminar, Omni-directional flow of unbound kinetic energy from every Planck ordinate point in space. The UniKEF energy flow was created at the time of the event currently considered the Big Bang but in UniKEF terms was more of a Big Rip.

It did not start from a singularity where all energy and matter in the universe was contained in zero volume according to the Big Bang view but occurred over a surface area creating a larger volume eliminating the necessity of there being an inflationary period where space expanded faster than the speed of light (FTL). The current expansion of space is a continuation of the Big Rip process.

While not defined in physical terms this inception can be viewed as originating from a bifurcated nothingness. Where "N" is nothingness and "S" is something, our existence can be described as:

N-----------------> { (+ S) + (- S) } = 0

Where - S can be energy absorption creating gravity and apparent time flow by communicating change or events to every ordinate point sequentially.

Space is formed by flowing unbound energy while matter is bound energy or compacted space moving in harmonic standing waves of a relativistic orb or vortices. The compaction factor is thought to be linked to the mass-energy ratio described by $E = mc^2$.

So not only did we arise from nothing but collectively our current existence totals nothing. Hence, nothing was ever created and there is no need for a Creator. This may not be easy to grasp but certainly, it is mathematically sound and has a terminal conclusion unlike other theories of origin, which have unresolved issues.

As unbound energy, (space) flow penetrates through highly compacted bound energy (mass) it imparts momentum to the mass in the vector of the energy flow.

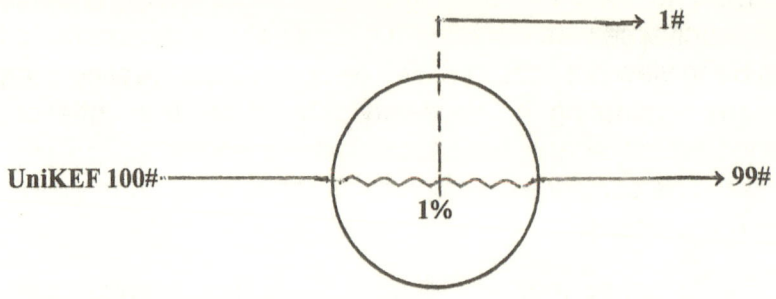

Single Body - Single Ray

Fig #6

In Fig #6 a UniKEF ray of energy, arbitrarily and by way of example only, is shown as having a value of 100# force and is flowing from the left to the right and is penetrating a spherical volume of mass shown as a circle. The mass provides a resistance to passage of 1% and that would cause the mass to receive a drag force of 1# toward the right. The UniKEF ray continues with only 99# force.

Zero Net Force

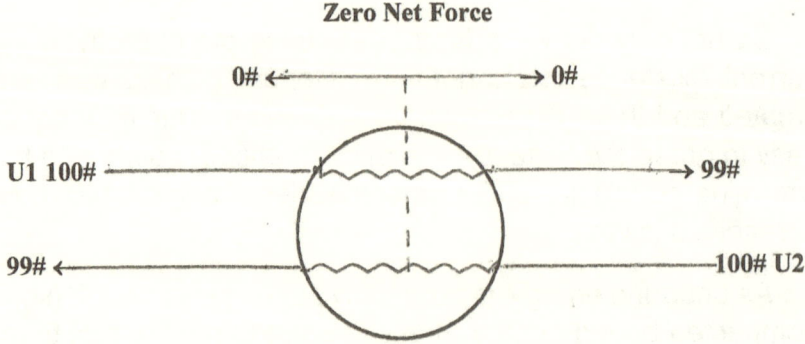

Fig #7a: Single Body - Opposing UniKEF Ray - Zero Net Motive Force.

In Fig #7a opposing UniKEF rays U1 & U2, are depicted. While they are actually on the same plane they are drawn separated for clarity of function. Note that the consequence of such opposing rays is to create a differential force field but results in a zero net motion of the body.

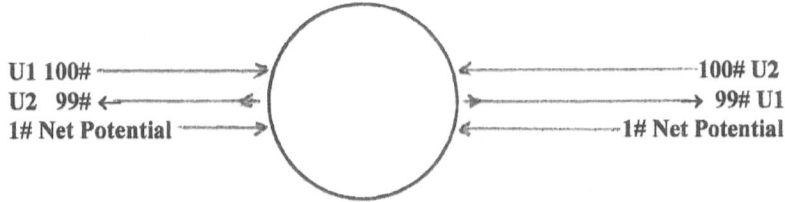

Fig #7b: Opposing Rays Net Force Field.

In Fig #7b, the opposing fields generate a net potential around the mass.

UniKEF considers this function to be substantially an elastic reaction but with some minor inelastic characteristics, a natural consequence would be production of heat within a gravitating body. This was the basis for the Priori Prediction about planet heating made in 1954.

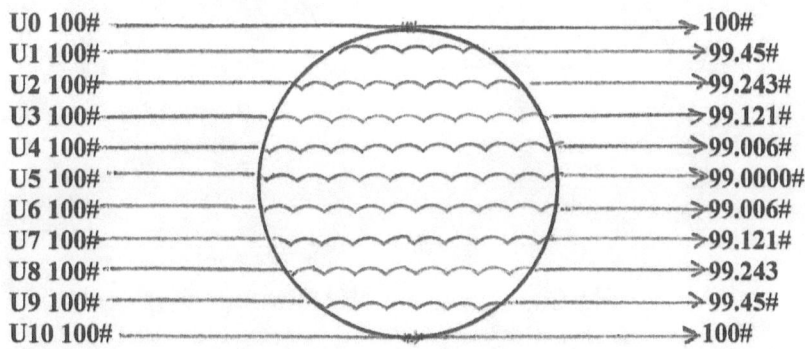

Fig 8 shows that rays penetrating a sphere have the most affect through the center. That is the amount of mass penetrated varies the amount of attenuation and drag force produced.

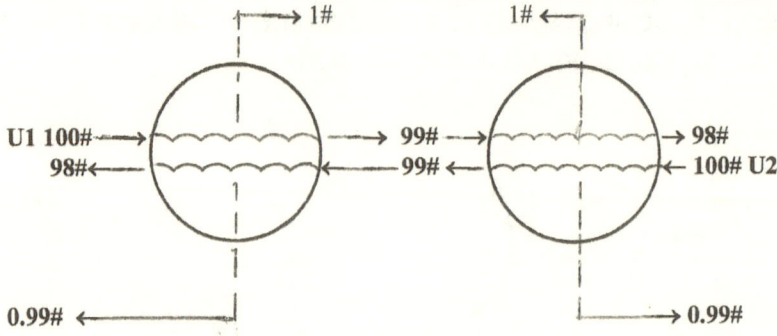

Fig 9a: Two Body - Opposing Rays

Fig 9a: Two Body - Opposing Rays.

Fig 9a shows a two-body system with opposing rays where the first mass is given greater momentum than the second penetrated mass in the vector of the UniKEF ray.

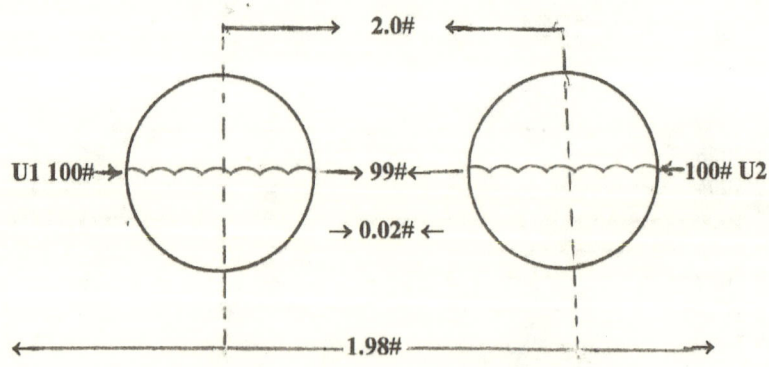

Fig 9b: Apparent Net Attraction

Fig 9b shows that the differential push forces of UniKEF attenuation create a net "Apparent" attraction.

For UniKEF to generate a gravitational force on a body the ray must penetrate two or more masses. Otherwise, it only produces heat and a gravitational potential around the one body mass.

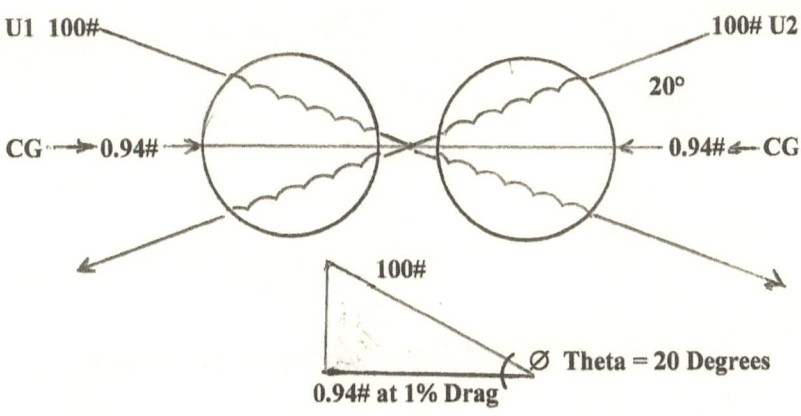

Fig #10 Effective angle of Penetration

CG is Center Line of Gravity

Fig #10 shows that the angle of penetration relative to the line of gravity through the center of masses has a trigonometric function. With a 100# UniKEF ray and 1%, attenuation, at an angle of 20 degrees to the line of gravity, produces a net apparent attraction of 0.0186#, not the 0.02# force at zero degrees along the line of gravity.

Force = 2*U * ~ * Cosine Theta

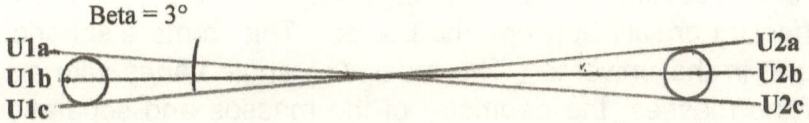

Fig #11: Source Cone at Distant Separation

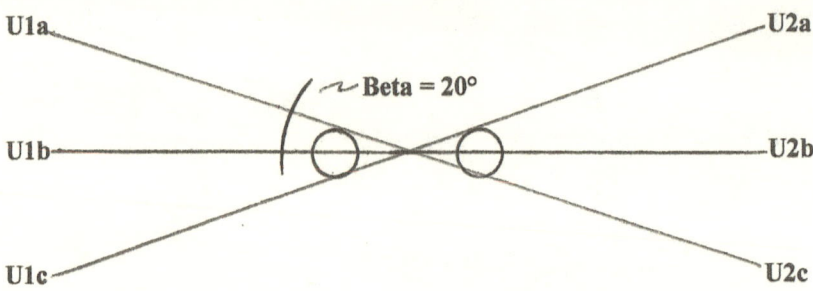

Fig #12:   Source Cone at Moderate Separation

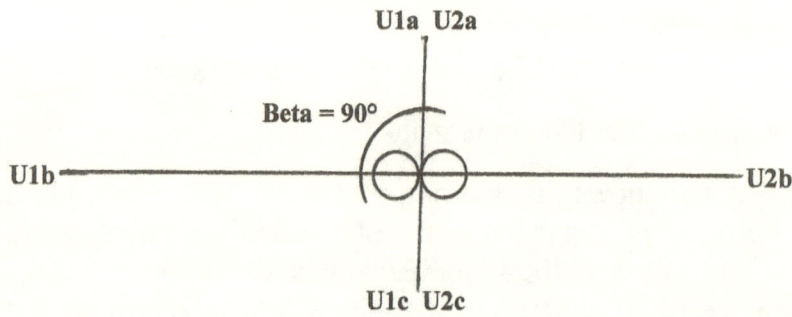

Fig #13: Source Cone at Surface Contact (2r Separation)

Fig #11, 12 & 13 show that for UniKEF to penetrate two masses there is an angle limit of the origin of sources (spatial ordinate points) from which energy may flow from the universe effecting gravity between the bodies. This forms a spherical cone in the universe. The maximum angle varies with size of the masses, the geometry of the masses and separation distance between the masses.

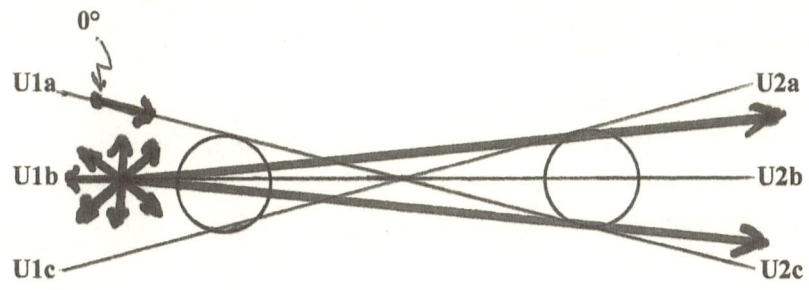

Fig #14: Dispersion Angle Limit Source Close to Bodies -
Bodies Close

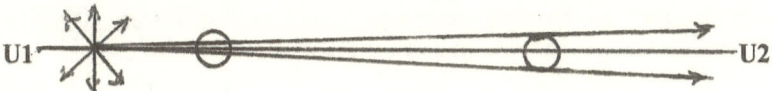

Fig #15: Dispersion Angle Limit Source Close but Small
Bodies separated more distant

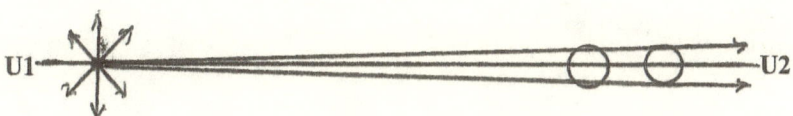

Fig #16: Dispersion Angle Bodies Close - Source at more
Distant Separation

Fig #14, 15 & 16 show non-parallel rays from an ordinate
point within the source cone along the line of gravity. The
maximum dispersion angle varies with the distance of source
from masses, mass sizes and separation between the
masses.

It also varies for origins from within the source cone that
are off to the side from the line of gravity. Sources along the

tangent line of the source cone have limited dispersion angle penetrations of both masses.

While the dispersion angles exist, they can generally be ignored since the distance of sources (radius of the universe) is so large versus the diameter of gravitating masses the statistical accumulative effective dispersion angle is zero degrees.

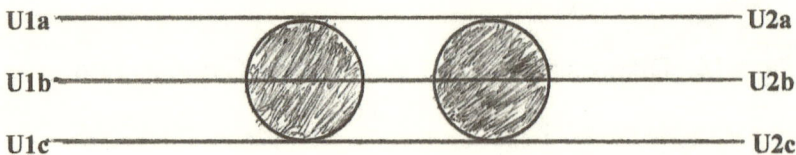

Fig #17: Volume Penetration of Parallel Rays at Zero Degrees.

Fig #18: Volume Penetration of Parallel Rays at Moderate Degrees Angle.

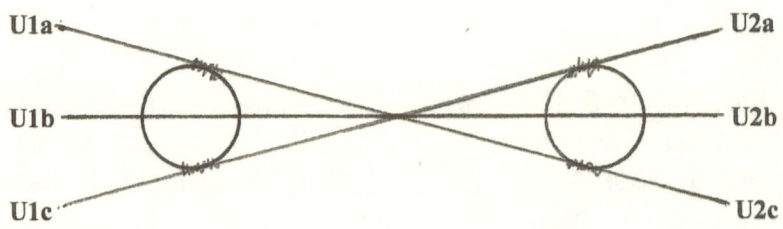

Fig #19: Volume Penetration of Parallel Rays at Maximum Degrees Angle.

Figures 17, 18 & 19 illustrate the penetration volume changes with angle.

A calculus integration of the volumes of penetration at all angles from zero degrees to the maximum source cone angle, compensated for the trigonometric angle to the line of gravity, is considered a UniKEF Pseudo Volume" or UPV and was found to be inverse square relative to the distance of separation.

The UPV is not the actual volume of the sphere but is the volume as seen by UniKEF from ordinate points within the source cone; including the compensation for the trigonometry function to the drag force at the angle of penetration.

Where d is mass density and PM is "Pseudo Mass". PM = d * UPV. This is not the actual mass. It is the integrated mass as viewed from the UniKEF source cone.

Where "U" is UniKEF energy and " ~ " is an attenuation or absorption coefficient, $F = U * \sim * PM$ is the momentum delivered to the mass. The product of F1 * F2 would be required to yield the net gravity force.

This is the product of an energy exchange and not the product of m1 * m2 or a mass squared term. To achieve a linear mass function the attenuation must vary with the UniKEF field strength at penetration. UniKEF must function as dynamic impedance rather than a mere passive resistance.

This is logical and is similar to the dynamic impedance where voltage produces a current flow and the current flow generates a counter emf. In that manner if you double UniKEF Field you quadruple the drag.

Where a 100# ray at 1% would cause a 1# drag and a 70.7# field yielding 0.707# drag at 1% it would instead produce 0.707 * 0.707 = 0.5# drag or only be 0.5% attenuation. Where "k" is the scaling factor gravity then becomes:

$F = U * \sim^k * (PM1 + PM2)$

This makes gravity logical as a function of total mass and not mass squared. After all what is mass squared physically?

A dynamic energy field can have a square function and be logical. That is the equivalent of the impulse of two cars hitting head on at "v". Such a collision force is the square of a car hitting a brick wall at the same "v".

Mass squared is a non-descript entity and is an illogical term physically. The fact that gravity is proportional to mass squared actually mandates that gravity is the product of an energy exchange in mass and not a function of mass per se.

The current gravitational constant is $G = 6.67E-11 N*m^2/kg^2$, by way of example only, may be divided into $U = 2.2233E60$ $N*m/s$ and $\sim = 3.0E-71 \ m*s/kg^2$.

The magnitude ratio of U versus $\sim$ must be such that it accommodates observed Black Hole gravity to gravity at atomic scales. Once U is totally blocked UniKEF predicts gravity will no longer increase with more mass and that one day we will discover that there is a deviation between mass and gravity, which will appear in super massive bodies.

To accommodate the $F = U * \sim^k * (PM1 + PM2)$ format change the kg^2 term to kg. In UniKEF, the kg is a function of PM and a new G or UG must be computed using:

UG = G * Actual Volume / UPV, an actual volume comparison conversion of G to UG.

When the energy momentum transfer is made to a mass that is restrained from accelerating, the vector push results in the production and a sensation of apparent weight.

Where distance not moved in a given period becomes - D. Work = F * (- D) = - Work.

Power then becomes - work / time = - P or gravity is viewed as energy absorption.

Current physics sees gravity as only conservation of potential energy. That is energy expended by raising a weight is Gravitational Potential Energy (GPE) = mgh = mass * gravity * height.

However, in reality GPE is being conserved by an active input by UniKEF. Gravity continues to conserve GPE in the form of force or weight for as long as the mass is restrained.

Once the restraint is, removed gravity converts GPE into Kinetic Energy (KE) by accelerated free-fall of the mass.

The following very general case was considered for the priori prediction about a gravity shadow during an eclipse:

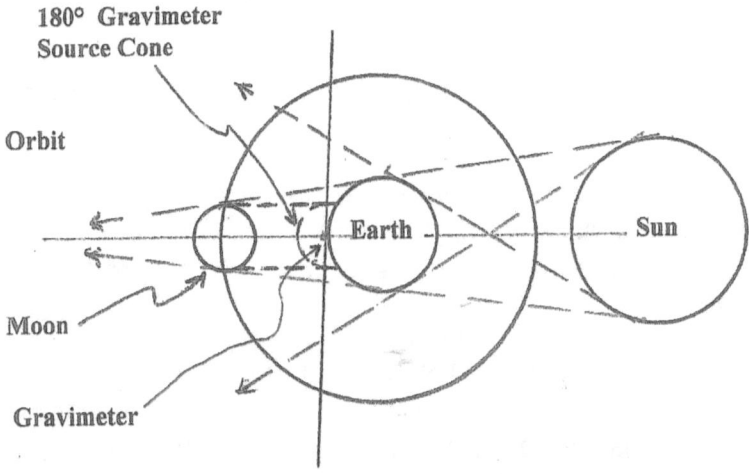

Eclipse of the Moon Gravity Pertabation
Fig #20

Fig #20 shows a shadowing affect of the earth and moon from the perspective of a gravimeter on the earth's surface in the umbra of the lunar eclipse. Due to the distances involved, a simplified calculation involving the various spatial components before, during and after an eclipse was done using the following volumes:

The technique used a cylindrical volume between the moon and earth that was 240,000 miles long and 3,000 miles in diameter to match the moon size and distance from earth.

$Vc = Pi * r^2 * L = 3.14159 * (1,500 \text{ miles })^2 * 240,000 \text{ miles} = 6.5E12 \text{ miles}^3$

I also used a volume of the source cone of the gravimeter on the earth's surface out to the orbit of the moon, which is a hemisphere.

$Vh = \{ (4/3) * Pi * r^3 \} / 2 = \{ (4/3) * 3.14159 * (240,000 \text{miles})^3 \} / 2 = 1E17 \text{ miles}^3$. Squared to yield the product of forces:

$Vc^2 = 4.2E25 \text{ miles}^3$

$Vh^2 = 1.0E34 \text{ miles}^3$

The ratio $Vc^2 / Vh^2 = 4.2E25 \text{ miles}^3 / 1.0E34 \text{ miles}^3 = 4.2E-9$.

This represents the predicted deviation in gravity as the gravimeter passes through different UniKEF field strengths caused by attenuation of the energy in the production of gravity.

During an eclipse of the moon in Norway, the Institut For Angewandte Geodeasie, Frankfurt a.M. den Hausanschlufs, Germany, discovered a gravity shadow. In their report "IfAG Mitt. No. 16, DKG Reihe B No. 234, they measured a deviation of 4.2E-6 Gal out of a total G field of 980 Gal.

A perturbation of 4.2E-6 Gal / 980 Gal = 4.28E-9!

In 1971, the institute granted me permission to publish the UniKEF manuscript and prediction referencing the correlation to their findings. See Appendix "B".

So far, we have only considered UniKEF incoming from the external source cone. Every spatial ordinate point between gravitating bodies produces repulsive UniKEF flow.

Therefore as the separation between masses becomes vast on a cosmic scale where the volume of sources repelling the bodies exceeds the external sources gravitating the bodies they begin to accelerate apart. The further apart the greater the repulsive forces.

This fact completes the series of transitions observed of gravity functions over the scale of the universe.

1 - Local Inverse Square

2 - Enhanced Gravity at Galactic distance scales, without MOND or Dark Matter

3 - Accelerated Expansion of the Universe on Cosmic Scales without Dark Energy.

UniKEF considers distance as a matter of "Quantitative" energy separation.   That led to the priori prediction that distance along the line of gravity is not a common quantity to all observers but would vary in proportion to the masses of the observer and observed long the line of gravity.

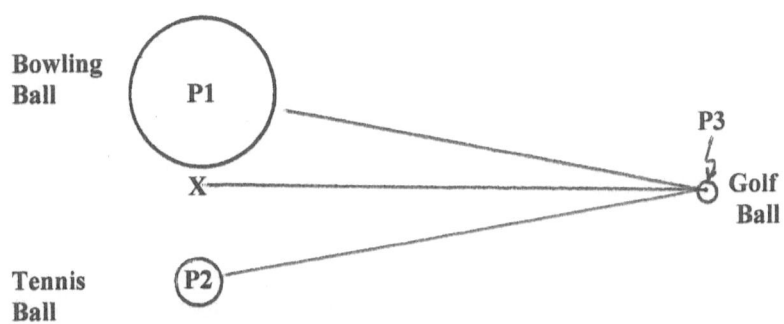

Distance According to a Remote Observer X
Fig #21a

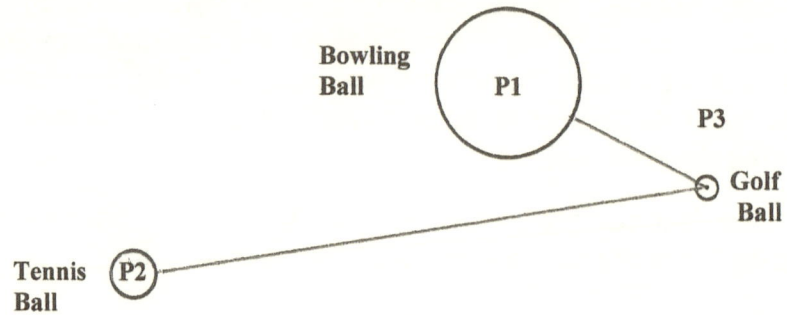

Distance between Masses
Fig #21b

Fig's 21a & 21b demonstrate the concept of variable distance as a function of the mass of the respective observers. This priori comes very close (similar) to the view expressed in General Relativity where space (distance) is curved or distorted by mass.

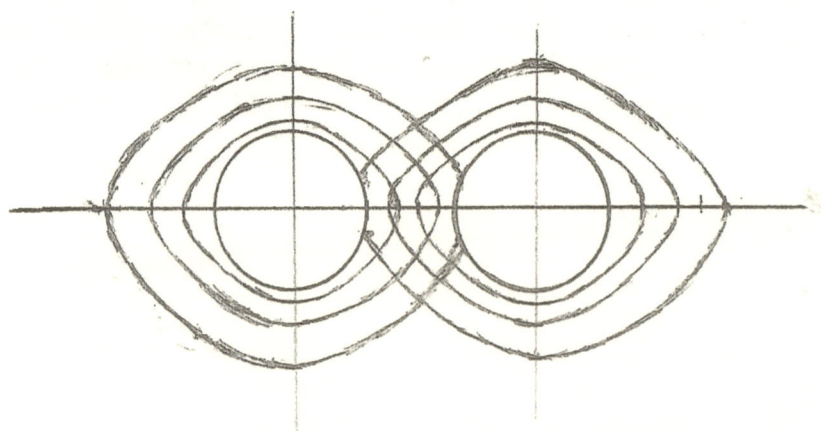

Fig 22: Spatial Dimensional Contours According to UniKEF

The spatial distortions shown in Fig 22 are difficult to measure since as the distance shrinks so does the length of rulers. Such that if distance goes from 12 feet to 9 feet your ruler will go from 12 inches to 9 inches and you will still measure 12 feet.

This makes space curvature virtually invisible and only the consequences of its existence are seen. Whereas General Relativity has space curvature causing gravity, UniKEF has gravity causing space curvature.

Such distance variation may one day become important when humankind begins interstellar missions. Gravity maps between super-massive Black Holes will become autobhans and Black Hole maps will become interstates shortening travel time between points in space.

The difference may appear moot but it is not. Curved space provides no cause to produce gravity. It is relegated to just being a mathematical description and not a physical description of cause.

In UniKEF, unlike Newtonian or Relativity theories, gravity has a cause and gravity generates the curved space and expansion of the universe.

Space is created by unbound energy. Gravity is the consequence of unbound energy attenuation by mass. Distance is the quantity of energy separation (space) at your inertial energy level universally and it is diminished by production of gravity.

Active bodies are converting bound energy into unbound radiant energy creating local increases in repulsive forces. The accumulative affect of the UniKEF within the source cones over rides the collective repulsion and the net result is that matter tends to coagulate or clump into solar systems, galaxies, galactic clusters and super clusters of matter.

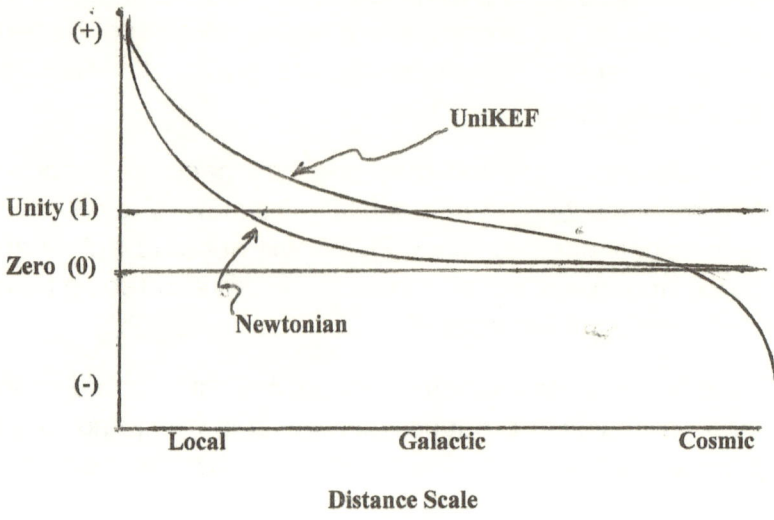

Fig #23 Comparative Gravity Functions

This shows the affect of having an active external energy produced gravity versus a purely mathematical internal non-energy Newtonian prediction. Where Newtonian gravity or General Relativity fail to account for star motion at the galactic scale and do not explain universal expansion, nor why gravitational force is a function of mass squared and not directly proportional to mass, UniKEF generally offers a physical cause over the entire spectrum of current observation.

Where universal flowing energy through mass produces gravity, accelerating mass through the UniKEF medium (F = ma) produces our sense of inertia hence it's mass.

You can see that UniKEF is inverse square locally but at galactic scales is much stronger than predicted by the Newtonian view. At cosmic distances, it becomes repulsive causing the accelerating expansion of the universe.

By scaling the universal UniKEF integrations to match the observed flattening of gravity at galactic scales and expansion at cosmic distances, a new calculation of the dimension of the universe can be established. The universe is predicted to be much larger than the currently observed matter universe.

For UniKEF gravity to function the universe MUST BE FINITE. The universe is viewed as an expanding, generally spherical volume, from a Big Rip type event. If the universe were infinite then the U term would be infinite which would produce infinite gravity and could not change with distance of separation.

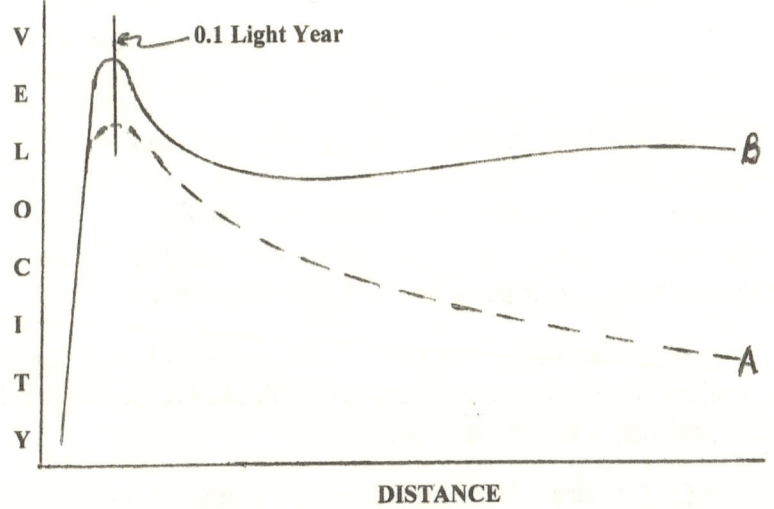

Fig 24: Expected (A) and Observed (B) star velocities as a function of distance from the center of a galaxy.

The importance of the deviation between current gravity theory and observation is best demonstrated by the above graph. Expected star motion is only predicted correctly out to about 0.1 light year or 5.875E11 miles.

That is 160 times the distance to the former planet Pluto or 10% of the way to the Ort Cloud that surrounds our solar system and is only 2.5% of the way to our nearest star neighbor - Alpha Centauri.

That means current gravity theory is virtually worthless beyond our immediate planets.

With the universe being finite then as the separation distance between bodies increases the proportion of internal repulsive sources increases relative to the external sources in the source cone.

At cosmic distance scales, the repulsive forces begin to exceed the external sources and masses begin to become repelled. The further out the greater the repulsion which resulted in the priori prediction that the universe expansion would be found to be accelerating. A feature now virtually considered fact.

Collectively the UniKEF energy flow associated with an ongoing creation (Big Rip Expansion) of space, generates gravity in masses in the interior regions of the universe but causes accelerated recession and expansion of mass distribution with a balance between gravity and F = ma in the expansion.

An energy exchange causes events and events result in our perception of time flow but time is an artifact of an energetic 3D space and not a 4th dimension.

UniKEF Gravity differs from LaSage and other push gravity concepts because they only considered external push forces. UniKEF envisions energy flowing from all spatial Planck ordinate points that includes sources of outward push from points between gravitating bodies.

Flow from between gravitating bodies actually pushes them apart reducing the net inward push. However, as Fig 23 shows locally gravity force is the normal inverse square relationship to distance of separation. As they separate the source cone diminishes more slowly than the $1/r^2$ rate of Newton and yields greater gravity at galactic scales.

As distance continues to increase, the source cone becomes a virtual cylinder and repulsive flow from in between bodies gets larger while the external gravitating force grows weaker.

Since normal gravity is based on a point source or center of mass concept it's mathematics trends toward zero at infinity. The UniKEF function trends toward unity at infinity as depicted in Fig #17 where current gravity (excluding quantum gravity) has three ranges. Local inverse square, galactic where Dark Matter is required to increase gravity and cosmic which requires Dark Energy or anti-gravity to accelerate the expansion of the universe. UniKEF is one cause covering the entire range of observed gravity functions.

A Planck length is 1.616252E-35 Meters and a planck time interval is the amount of time it takes light to travel that distance which is 5.39129E-44 Seconds. Therefore, the flow velocity could double 185,000,000,000,000,000,000,000,000,000,000 ,000,000,000,000 times every second.

Since these numbers exceed the capacity of everyday, computers to put doubling this often per second into perspective consider starting with 1 cent. Double it every second and after merely 27 seconds you will have $1,342,177.28!

Likewise if you increase planck length to 1.616252m by dropping the E-35 term and increase planck time to 5.39129E-9 by multiplying it by E35 and then compute how long it takes the UniKEF wave front to reach v = c.

You will find that it exceeds 3E8m/s after only 11 cycles. That would only take 5.9E-8 seconds or 1/17,000,000ths of a second. The wave would have traveled 5.8E16m or 3.6E13 miles.

If you now scale these numbers back by E35 then UniKEF reaches, the speed of light in 5.9E-43 seconds or in one second the velocity would be 1.7E42 times the speed of light! Distance to reaching v = c would be 5.8E-19m or 2.28E-17inches. That means the wave front reaches an incredible

43,804,317,350,000,000 times the speed of light in just 1 inch!

The UniKEF wave could reach across the observable universe 2.5E32 times every second (2,500,000,000,000,000 ,000,000,000,000,000,000 times/s). Einstein's spooky action

at a distance is resolved.

These numbers are based on each ordinate point effusing energy every Planck time interval. That is an unknown, it may be that such effusion is spuratic or like statistical decay, intermittent, etc.

However, should that be the case there happens to be 8.78E60 Planck lengths in the radius of the universe that is 8.78 trillion, trillion, trillion, trillion, trillion Plancks. If every Planck length ordinate point effuses once in 80 million billion billion years, the wave velocity still would double 245 million times per second!

CHAPTER 7

UniKEF Calculus

Dr Edward Allard, Physicist, performed the following calculus presentation for me   while I was in nuclear power school in 1965.

For the benefit of current readers, I point out that other mathematicians have reviewed this work and commented that they took procedural differences with the presentation but that they do not affect the conclusion, which was that the UniKEF view locally is inverse square in form.

In addition, I have found it necessary for publication to transcribe his hand written copy into type and may have inadvertently made omissions or other errors.

Dr Allard's work only considered the gravitating masses as viewed by energy from the external source cones and does not compute the internal out push with increased separation. Therefore, it is only valid for the local inverse square gravity regime.

The results are the geometric component and not actual gravity force.   To get the force the geometry must be multiplied by the UniKEF field "U" energy magnitude and the "~" absorption or attenuation terms.

CALCULUS DERIVATIONS

Technical Evaluation by Dr Edward Allard, Physicists

According to Newton's Law of Attraction, two circles of the same radius and density attract each other with a force given by:

$$F = \frac{K\ m_1\ m_2}{r^2} = \frac{K(\sigma\pi R^2)\ (\sigma\pi R^2)}{r^2}$$

If the separation distance is R

$$F = \frac{K(\sigma\pi R^2)\ (\sigma\pi\ R^2)}{R^2} = K\sigma^2\ \pi^2\ R^2$$

Let us compare this result with the Uni-K-E-F theorem.

Consider two circles with equal radius R, and equal density, separated by R. This assumption is used for mathematical convenience.

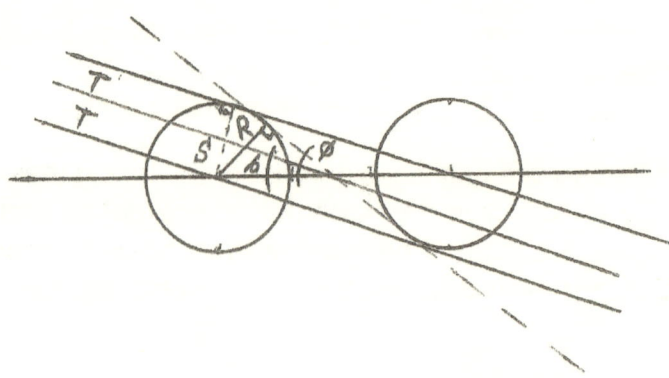

$$\text{Sin}\ \varnothing\ = \frac{R}{\frac{3R}{2}} = \frac{2}{3} \qquad S = \frac{3R}{2}\ \text{Sin}\ 6 \qquad T = R - S = R - \frac{3R}{2}\ \text{Sin}\ 6$$

Fig 25: Two gravitating bodies at R surface separation

The sum of the total areas will be the area covered by 2T times 2, since the problem is symmetrical.

Find the area in the circle.

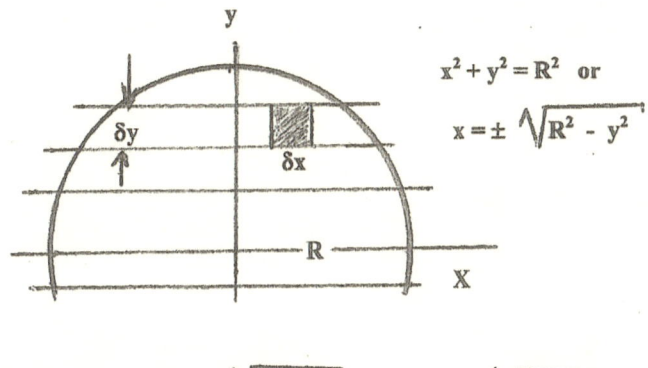

$$x^2 + y^2 = R^2 \quad \text{or}$$

$$x = \pm \sqrt{R^2 - y^2}$$

$$A = \iint \delta y\, \delta x = 2 \int_{0}^{R} \delta y \int_{0}^{\sqrt{R^2 - y^2}} \delta x + 2 \int_{R-2T}^{0} \delta y \int_{0}^{\sqrt{R^2 - y^2}} \delta x$$

Fig 26: Area of integration

Notice that this area is the same as the area we are trying to calculate. Just rotate the coordinates by -6 degrees.

$$A = 2 \int_{0}^{R} \delta y \sqrt{R^2 - y^2} + 2 \int_{R-2T}^{0} \delta y \sqrt{R^2 - y^2}$$

Integrating:

$$A = 2 * \tfrac{1}{2} \left[y\sqrt{R^2 - y^2} + R^2\ Sin^{-1}(y/R) \right]_{0}^{R} + \tfrac{1}{2} * 2 \left[y\sqrt{R^2 - y^2} + R^2\ Sin\ y/R \right]_{R-2}^{0}$$

$$= \left[R^2\ Sin^{-1} 1 - R^2\ Sin^{-1} 0 \right] + \left[-(R - 2RT)\sqrt{R^2 - (R - 2T)^2} - R^2\ Sin^{-1} - R - \frac{2T}{R} \right]$$

$$= \pi R^2 - 2R^2 (3\ Sin\ 6 - 1)\sqrt{3/2\ Sin\ 6 - 9/4\ Sin^2\ 6} - R^2\ Sin^{-1}(3\ Sin\ 6 - 1)$$

The sum of two areas multiplied by Cos ϐ is

$$\Sigma A \cos \theta = \left[\pi R^2 - 4R^2 \left(3 \sin \theta - 1 \right) \sqrt{\frac{3}{2} \sin - \frac{9 \sin^2}{4} \sin^{2^3}} - 2 R^2 \ \sin^{-1} \right.$$

$$\left. (3 \ \sin \theta - 1) \right| \cos \theta$$

ϐ goes from -Ø to Ø where $\emptyset - \sin^{-1} \dfrac{2}{3}$

① ②

$$\overrightarrow{V} \sim 2 \int_0^\emptyset \pi R^2 \ \cos \theta \ \delta \theta \ 8R^2 \int_0^\emptyset (3 \sin \theta - 1) \sqrt{\frac{3}{2} \sin \theta - \frac{9 \sin^2 \theta}{4}} - 4 \Bigg| R^2 \int_0^{\cos 6} {}^{\delta 6 \ \emptyset}$$

③

$$\left[\sin^{-1} (3 \sin \theta - 1) \right]^{\cos \theta} \delta \theta$$

The second terms in this, integral vanishes.

$$V \sim 2 \int_0^\emptyset \pi R^2 \ \cos \theta \ \delta \theta - 4 R^2 \int_0^\emptyset \sin^{-1} (3 \sin - 1) \ \delta \theta$$

$$\sim 2\pi R^2 - \frac{2 R^2}{3} \left[\frac{\pi}{2} - 1 \right] = \frac{2 R^2}{3} \left(\frac{\pi + 1}{2} \right)$$

If the separation is given by some distance D times R

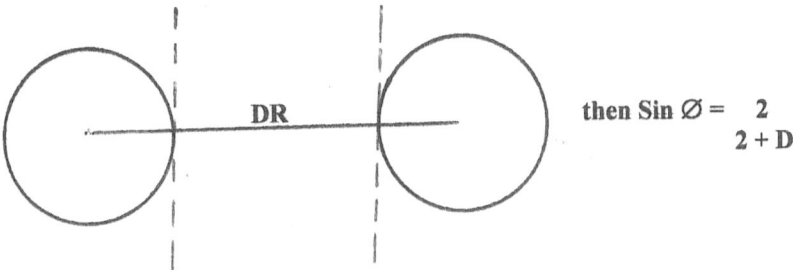

DR   then Sin $\emptyset = \dfrac{2}{2+D}$

Fig 27: Separation by D*R

This more general problem gives

$$\sim 2\int_0^\emptyset \pi \, R^2 \, \text{Cos} \, \text{δ} \, \text{δ} \, \text{δ} \, - 2R^2 \int_0^\emptyset \left[\text{Sin}^{-1}\left[(2+D) \, \text{Sin} -1\right] \text{δ} \, \text{δ} \, \text{Cos} \text{δ} - \int_0^\emptyset 4R^2\left[(2+D) \, \text{Sin} \, \text{δ} - 1\right] \right.$$

$$\sqrt{(1+\underline{D})^2 \, \text{Sin}^2 \, \text{δ} -(1+\underline{D}) \, \text{Sin} \, \text{δ}} \; \text{Cos} \, \text{δ} \; \text{δ}$$
$$\qquad\qquad 2 \qquad\qquad\quad 2$$

The third term in this integral vanishes giving

$$\sim 2\int_0^\emptyset \pi R^2 \, \text{Cos} \, 6 \, \text{δ} \, 6 - 2R^2 \int_0^{\emptyset'} \text{Sin}^{-1}\left[(2+D)\, \text{Sin} \, 6 - 1\right] \text{Cos} \, 6 \, \text{δ} \, 6 \; \sim \; \pi R^2\left[\dfrac{2}{2+D}\right] - \dfrac{2R^2}{2+D}\left[\dfrac{\pi-1}{2}\right]$$

$$\sim \boxed{\dfrac{2\,R^2\,(\pi+1)}{2+D\quad 2}}$$

This has the form of the potential between two circles. The derivative along the line joining the centers of the circles, gives the force between the two circles.

$$\boxed{F \sim 2\,R^2\,\dfrac{(\pi+1)}{2\quad D^2}}$$

$$-4R^2 \int_0^\emptyset (-1)\sqrt{\frac{(1+D)}{2}x - \frac{(1+D)^2}{R}x^2}\ \delta X \qquad -4R^2 \int_0^\emptyset (2+D)\,x\sqrt{\frac{(1+D)}{2}x - \frac{(1+D)^2}{2}x^2}$$

$$-4R^2 \ (2+D) * \left\{ \frac{X\sqrt{X}}{3C} - \frac{b}{2c}\left[\frac{(2cx+b)\sqrt{x}}{4c} + \tfrac{1}{2}K * \frac{1}{\sqrt{-c}}\ \mathrm{Sin}^{-1}\frac{(-2cx-b)}{\sqrt{b^2-4ac}} \right] \right\}$$

$$+4R^2 \left\{ \frac{(2cx+1)\sqrt{X}}{4c} + \tfrac{1}{2}K\left[\frac{1}{\sqrt{-c}}\ \mathrm{Sin}^{-1}\frac{(-2cx-b)}{\sqrt{b^2-4ac}} \right] \right\}$$

$$-4R^2 \ (2+D) * \frac{X\sqrt{X}}{3C} - \frac{b}{2c}\left[\frac{(2cx+b)\sqrt{x}}{4c} \right] - \frac{b}{2c} * \ \tfrac{1}{2}K * \frac{1}{\sqrt{-c}}\ \mathrm{Sin}^{-1}$$

$$\frac{(-2cx-b)}{\sqrt{b^2-4ac}} - 4R^2 \left[\frac{X\sqrt{X}}{3c} \right](2+D) +\sqrt{X}\left[+ 4R^2(2+D)\ \frac{*\,b}{8c^2} + \frac{4R^2}{4c} \right](2cx+b)$$

$$+\ \mathrm{Sin}^{-1}\frac{(-2cx-b)}{\sqrt{b^2-4ac}}\left[4R^2\ (2+D)\ \frac{*\,b}{4cK\sqrt{c}} + \frac{4R^2}{2K}\sqrt{-c} \right]$$

Where: $b = 1 + \dfrac{D}{2}$ $\qquad c = -(1+\dfrac{D}{2})$ $\qquad \tfrac{1}{2}K = -\dfrac{b^2}{8c} = \dfrac{-(1+D)^2}{\dfrac{2}{-8(1+D)^2}} = 1/8$

$X = \mathrm{Sin}$ $\quad a = 0$ $\quad -c = 1 + \dfrac{D}{2}$

$$X = (1+\underline{D}) \, \text{Sin} \, \phi - (1+\underline{D})^2 \, \text{Sin}^2 \, \phi$$
$$2 \phantom{\text{Sin} \phi - }2$$

$$-4R^2\left[\frac{x\sqrt[4]{x}}{-3(1+\underline{D})^2}\right] \; (2+D) \, \sqrt[4]{x}\left[-2(1+\underline{D})^2 \, x + (1+\underline{D})\right] \left[\frac{4R^2 \, (2+D) \; * \; (1+\underline{D})}{8(1+\underline{D})^4} \; + \right.$$

$$\left.\frac{4R^2}{-4(1+\underline{D})^2} \; + \; \text{Sin}^{-1}\left[\frac{2(1+\underline{D})^2 \, x - (1+\underline{D})}{(1+\underline{D})}\right]\right]$$

$$\left[\frac{4R^2 \, (2+D) \; * \; (1+\underline{D})}{-2(1+\underline{D})^2} \quad * \quad \frac{1}{8(1+\underline{D})} \; + \; \frac{4R^2}{8(1+\underline{D})}\right] \qquad 2+D = 2(1+\underline{D})$$

$$\frac{8R^2 \, x\sqrt[4]{x}}{3(1+\underline{D})} \; + \sqrt[4]{x}\left[-2(1+\underline{D})^2 \, x \; * \; (1+\underline{D})\right] \left[\frac{8R^2 \, (1+\underline{D}) \; * \; 1+\underline{D}}{8(1+\underline{D})^4} - \frac{R^2}{(1+\underline{D})^2}\right] \; +$$

$$\text{Sin}^{-1}\left[\frac{2(1+D) \, x - 1}{2}\right]\left[\frac{8R^2 \, (1+\underline{D})}{-16(1+\underline{D})^2} \; + \; \frac{4R^2}{8(1+\underline{D})}\right]$$

$$\frac{8R^2 \, x\sqrt[4]{x}}{3 \; (1+\underline{D})} \; + \sqrt[4]{x}\left[-2(1+\underline{D}) \, x + 1\right]\underbrace{\left[\frac{R^2}{(1+\underline{D})} \; - \; \frac{R^2}{(1+\underline{D})}\right]}_{0} \; +$$

$$\text{Sin}^{-1}\underbrace{\left[2(1+\underline{D}) \, x - 1\right]\left[\frac{-R^2}{2(1+\underline{D})} \; + \; \frac{R^2}{2(1+\underline{D})}\right]}_{0}$$

$$= \frac{8R}{3(1+\underline{D})} \; x\sqrt[4]{x}$$

$$\emptyset \;=\; Sin^{-}\;\frac{2}{2+D}$$

$$I_1 = 2\int_0^{\emptyset}\frac{\pi R^2}{2}\,Cos\emptyset\,\delta\emptyset \;=\; \pi R^2\,Sin\emptyset\;\Big]_0^{\emptyset}\;=\;\pi R^2\left[\frac{2}{2+D}\right]$$

$$\text{②}\quad \frac{-2R^2}{2+D}\left[x\,Sin^{-1}\;x+\sqrt{1-x^2}\;\right]_0^{\overbrace{(2+D\;Sin\;Sin^{-1}\;\frac{2}{2+D}\;-1)}^{1}}$$

$$\text{②}\quad \frac{-2R^2}{2+D}\left[x\,Sin^{-1}\;x+\sqrt{1-x^2}\;\right]_0^{1}\;=$$

$$\frac{-2R^2}{2+D}\left[Sin^{-1}\;1+0-0-0\right]\;=\;\frac{-2R^2}{2+D}\left[\frac{\pi}{2}-1\right]$$

$$I = \pi R^2\left[\frac{2}{2+D}\right]-\frac{2R^2}{2+D}\left[\frac{\pi}{2}-1\right]=\frac{2R^2}{2+D}\left[\pi-\frac{\pi}{2}+1\right]=\frac{2R^2}{2+D}\frac{(\pi-1)}{2}$$

$$X = \operatorname{Sin} \textit{6} \qquad x = \operatorname{Sin} \varnothing \;=\; \frac{2}{2+D} \;=\; \frac{1}{1+\dfrac{D}{2}}$$

$$\boxed{3} \;=\; \frac{8R^2}{3\dfrac{(1+D)}{2}} \left[\left[(1+\underline{D})\, x - \frac{(1+\underline{D})^2}{2}\, x^2\right] \quad (1+\underline{D})\, x - \frac{(1+\underline{D})^2}{2}\, x \right]_{0}^{\frac{2}{2+D}}$$

For $D = 2$

$$\boxed{3} \;=\; \frac{8\,R}{3*2}\left[(2x - 4x)\,\sqrt{2x - 4x^2}\,\right]_{0}^{\frac{1}{2}} \;=\; \frac{8\,R}{6}\Big[(1-1)\Big] \;\Rightarrow\; 0$$

For $D = 3$

$$\boxed{3} \;=\; \frac{8R^2}{3*\dfrac{(5)}{2}}\left[(5\,x - \frac{25}{4}\, x^2\,)\; \frac{5}{2}\,x - \frac{25}{4}\, x^2\right]_{0}^{\frac{2}{5}} \;\Rightarrow\; (5/2 * 2/5 - 25/4 * 4/25) \;\Rightarrow\; 0$$

CHAPTER 8

UniKEF Gravity Testing

UniKEF Gravity is a concept based on a universal kinetic energy flow creating space (unbound energy) which is attenuated and/or absorbed by mass (bound energy) or compacted space.

The concept was first formulated in 1954 but lay dormant due to lack of any physical evidence of such a field existing.

Recent work in the area of breaking down the "Vacuum or Voids" of space into constituent parts and finding vast stores of energy, as well as virtual particles coming into and out of existence, annihilating and releasing energy in a planck time interval of 5.39E-44 seconds, prompted a renewed effort to find a means of testing the UniKEF concept.

The test takes advantage of the fact that UniKEF gravity is not based upon the Center of Mass (COM) as it is in Newtonian and Relativistic Gravity theories. UniKEF gravity has a geometrical component that allows the theory to be tested.

A test was devised which measured the acceleration of gravity using a modified Cavendish Balance. The test mass was symmetrical with a port hole through its center and a remote port hole plug.

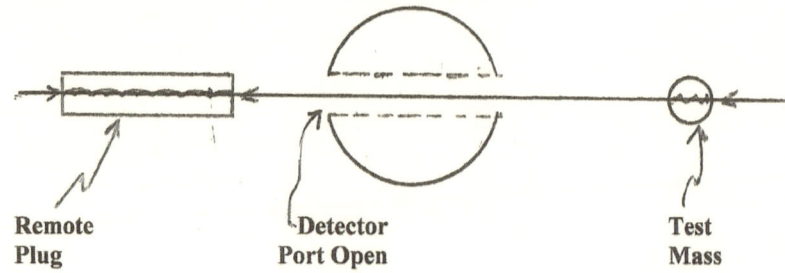

Fig 28a: Remote Port Plug in place Detector Port Open

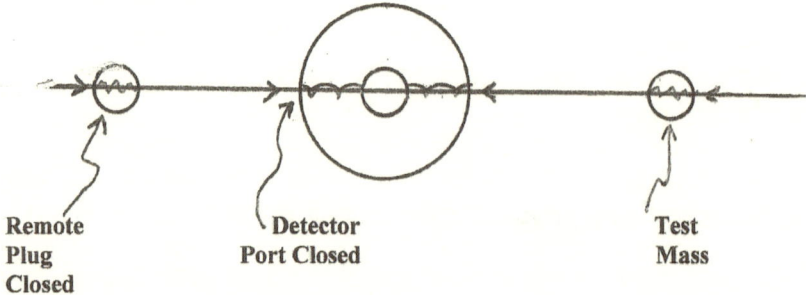

Fig 28b: Remote Plug Aligned Closed and Detector Port Closed

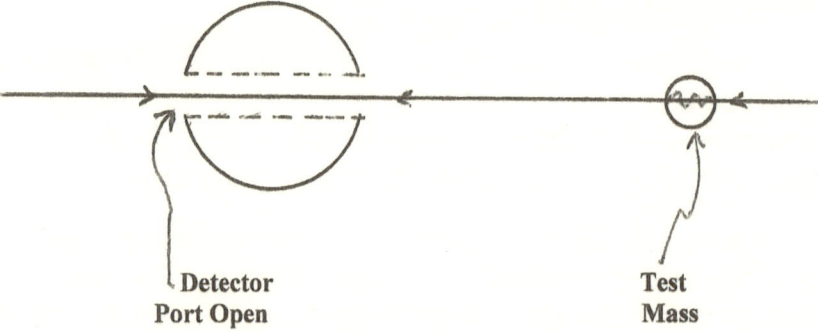

Fig 28c: Remote Port Plug Removed and Detector Port Open

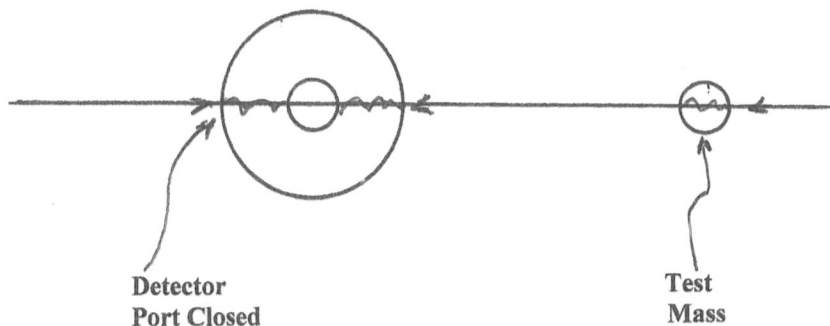

Detector
Port Closed

Test
Mass

Fig 28d: Remote Port Plug Removed and Detector Port
Closed

In Fig's 28a and 28b the use of a remote port plug mitigates the change in gravity when the port is rotated from open to closed and the remote port plug is also oriented with the detector port it is as though the detector mass is now solid with no port hole.

In Fig's 28c and 28d with the remote port plug removed there is measurable difference in the gravity response between the detector and test mass. When the port is open to the test, mass gravity is weaker even though the detector is symmetrical and rotating it does not alter either the mass, center of mass nor distance of separation.

The conventional Newtonian form of gravity formula $F = G$ * mass 1 * mass 2 / r^2 does not produce a different result since no component of the formula changes with the rotation of the detector port.

UniKEF gravity testing over several months consistently produced positive results, albeit somewhat erratic. However, the testing was statistically significant evidence of UniKEF streaming through the open port.

The test measured differences in gravity via the amount of time required for the test mass attached to the moveable torsion arm to go from the backstop to the Detector. The following test

Dan Keith McCoin

run data was made without a remote port plug on Monday, February 10, 2003 at 4:56 PM. All tests were conducted with a 4:00 minute delay between test runs.

Sample Test Data in Seconds Closure

Test #	Port Open	Port Closed	Differential
1	68.99	56.13	+12.86 (1)
2	55.34	45.92	+9.42
3	59.82	56.91	+2.91
4	59.34	48.13	+11.21
5	47.93	45.12	+2.81
6	42.12	55.11	-12.99 (1)
7	54.16	47.91	+6.25
8	68.54	60.31	+8.23
9	52.91	45.47	+7.44
10	57.65	52.12	+5.53
	56.68	51.31	+5.37
	Average	Average	Average

(1) High /Low Data Discarded

+5.357

Average

Calculations show that we were measuring response differentials based on F = ma where delta F was approximately 60 trillionths of a pound. So the test was not a direct force measurement but of response to force by timing the movable torsion arm test mass over a fixed distance from point of release between stops at the detector.

A slight reverse bias was set by an adjustable swivel link in the suspension fiber, to hold the torsion arm test mass back against an electrically grounded backstop when there was no detector mass present.

Such that any forward motion toward the detector was a direct function of the gravitating force of the Detector.

The operator, Mr. McClain, had to wear a ground cable to eliminate electrostatic charge affects generated by moving in and rotating the detector between tests. He also had to stand a sufficient distance away in the torsion arms general plane or his presence would affect the readings.

Temperature and humidity were monitored and regulated. There was an air curtain surrounding the test area to minimize any air current disturbances. The detector and torsion arm back stop were electrically grounded.

Numerous tests were run with 10 data points per session. Some data runs would be positive by 60 seconds. Others were positive by only 3 seconds. All were statistically positive.

We ultimately determined that the day to day differences in response was due to the positions of the sun and moon at the time of the tests.

We subsequently spent several weeks just tracking the torsion arm motion during the day, with no detector mass present and were able to plot it as a function of the position of our solar neighbors.

This was an unexpected result in that the arm was counter balanced. Testing showed that there was a slight difference in the mass of the counter weight on the arm and that was causing the arm to track the solar body mean forces. A second torsion arm was fabricated but resulted the same solar influence response.

The consequence was an influence in the response time as a function of the solar assist or retardation. Over the period of one hour of the testing periods, it was consistent but from day to day, the range of data drifted cyclically with solar body orientations.

While the testing was very convincing, it needs to be replicated and done with equipment that is more precise before one could claim it has proven anything. While it was not conclusive, it certainly was indicative that UniKEF might be a valid gravity concept.

The test detector pan can be filled with various materials to give different masses for testing baseline response purposes and to decrease test response times by increasing detector force and increasing the port open and port-closed differentials.

Mr. McClain builder and operator of the UniKEF Gravity Tester

UniKEF Detector Port Open to the Test Mass.

Remote Port Plug Aligned with Port Open - Detector Port Open.

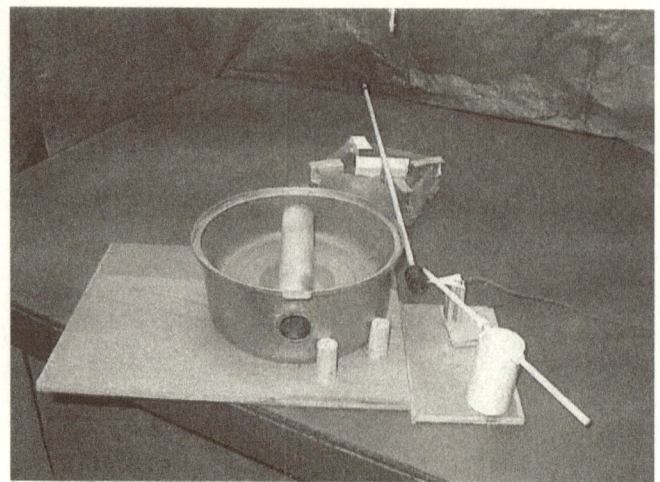

Side view of the Detector Port Closed. The electrical ground at the backstop, a whisker tool holding the test mass in place before release for timing, a water dampening brake and swiveling suspension line yoke are all visible in the background. The detector pan is filled with water.

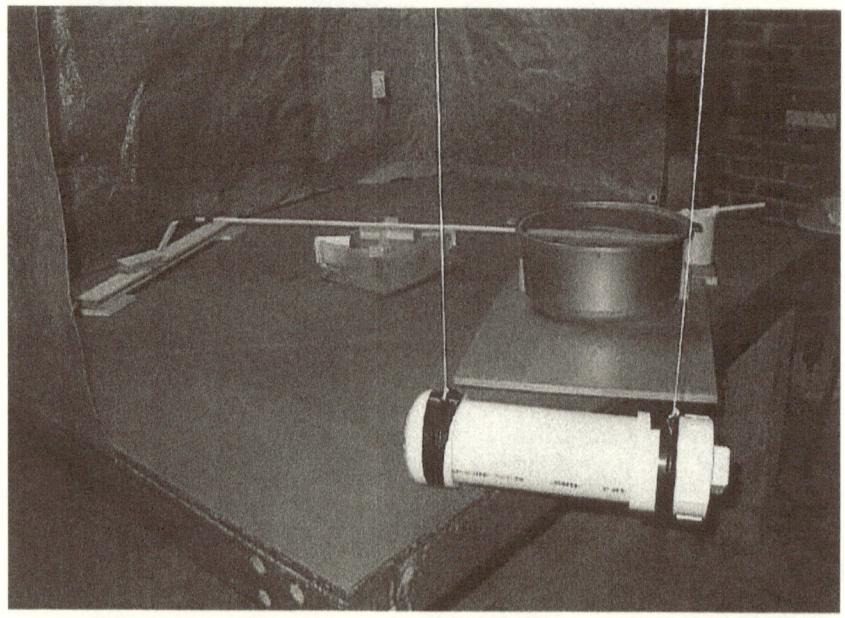

Remote Detector Port Plug Aligned Closed - Detector Port Closed.

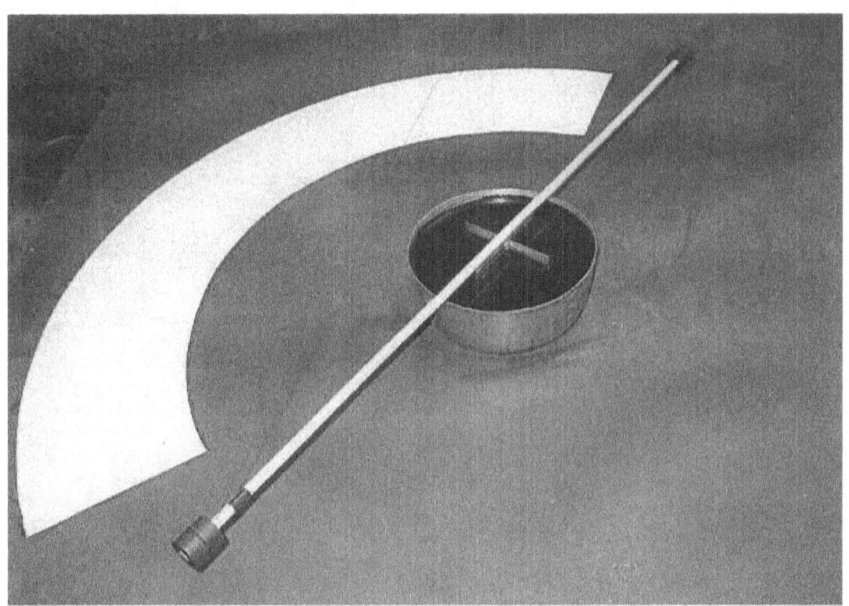

Start of a Daily Solar Test Cycle - 4/15/03 - Noon

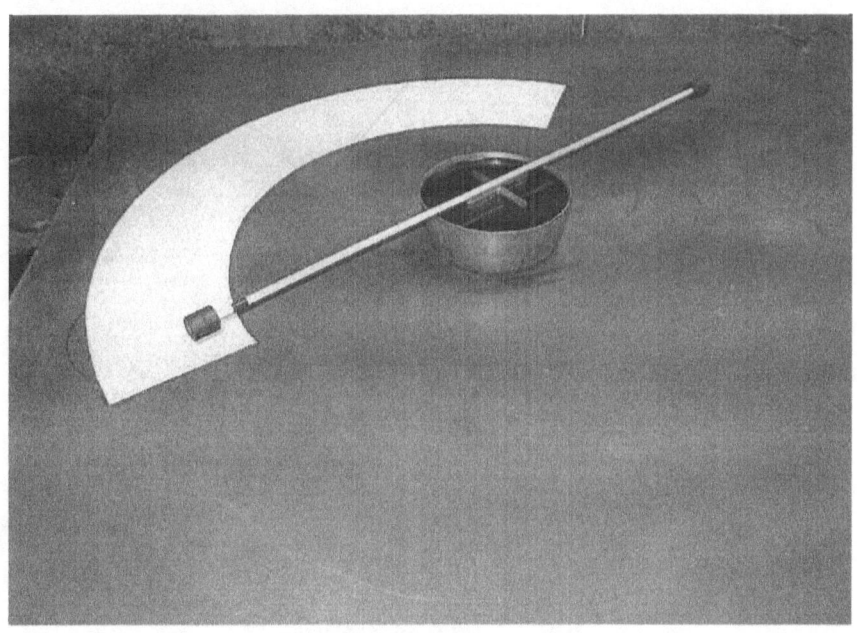

Solar Cycle at 2:30PM.

Solar Cycle at 4:00PM

Solar Cycle at 7:30PM.

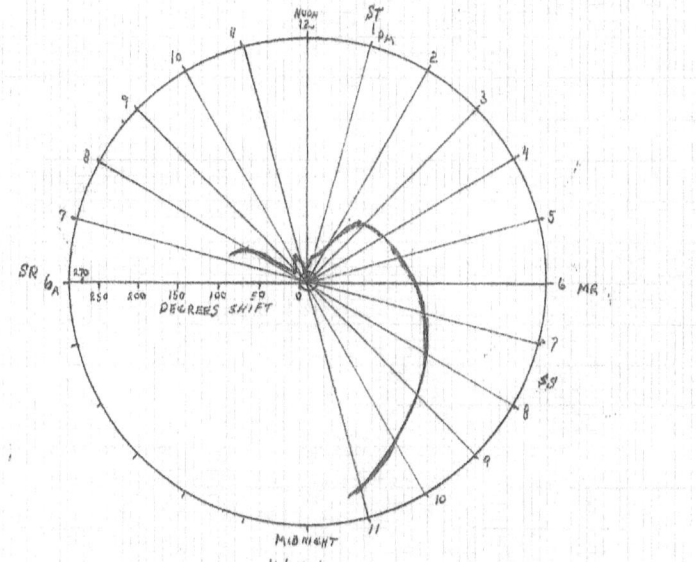

Chart of Solar Tracking Cycle.

SR is Sun Rise at 6:04AM. ST is Sun Transit around noon. MR is Moon rise and SS is Sun Set.

CHAPTER 9

Nuclear Force

From a UniKEF perspective, the nuclear force may well be a form of enhanced gravity. As has been indicated it is the attenuation of unbound energy by passage through compacted bound energy (mass) that results in a gravitational force.

Mass is viewed as spatial energy moving in tightly bound relativistic orbs or vortices forming a harmonic standing wave. Movement orthogonal to the line of gravity can be seen to increase the apparent density by intercepting a greater amount of UniKEF in proportion to the comparative velocity of motion versus the energy (velocity) of the UniKEF Field.

The following analogy using an upper hopper filled with BB's, a control gate, a moving screen and lower catch hopper; demonstrates the principle involved.

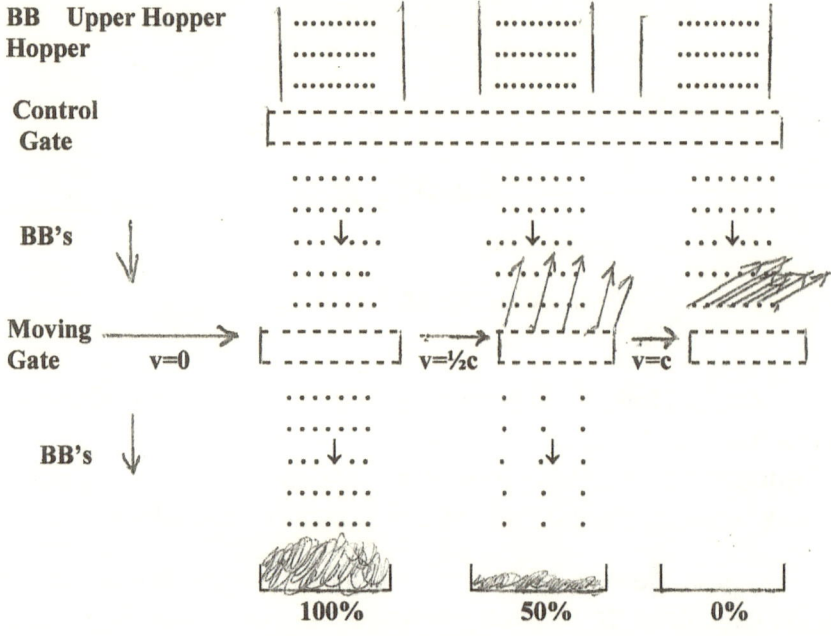

Fig # 30

The nucleus of an atom is generally 1E-15m while the atom is 1E-10m. That is the nucleus is only 1/100,000th of the size of the atom. The volume is 4/3*pi*r³ such that the nucleus is only 1/1,000,000,000,000,000th of the volume of an atom. Further, the nucleus is composed of neutrons and protons and they are composed of quarks. Today there is theory that quarks are made of the smaller components called strings.

In between these components is nothing but space. All material that we experience in everyday life is mostly just space and very little mass. Black Holes vary in density but many exceed 5E12 pounds per cubic inch because they have crushed out the normal space in an atom. That is 5 Trillion pounds/inch³.

Putting these numbers into something a little more understandable a sphere the size of a dot "." , weighs 41,000,000 pounds, as much as 16,000 full sized automobiles.

Nuclear forces are 1E35 times as strong as everyday gravity. That is 100 million, billion, billion billion times stronger. Keeping in mind that the actual physical mass is only a minor part of the atom it is not difficult to envision Black Hole strength gravity between super dense sub-atomic components.

The range between super strong attraction and super strong repulsion suggests a very small relativistic moving core with a lot of space between components. You can envision an atom as being a miniature solar system (Bohr view) where locally you have gravitational attraction but some distance away it will become repulsive.

The most likely scenario is that there is an enhanced gravity attraction at the sub-atomic level but it is countered by an enhanced coulomb type repulsion resulting in the abrupt nuclear force transition.

Put more into the atomic type structure envision a tube with BB's dropping through a rotating blade. The faster the blade turns the more BBs the blade intercepts. The rotating or vibrating sub-atomic components provides an increased intercept of the UniKEF and causes increased gravity between components.

On the macroscopic scale, gravity is based on the collective drag/volume such that it is like putting 1 ounce of pressure on a needle and creating 1,000 pounds/inch squared force on the point.

CHAPTER 10

Distance & Refraction

Distance:

In General Relativity (GR) "Depth of Field" describes the amount of time-space curvature and impacts time dilation. It is depicted as a gravity well or curvature of space around a massive object.

Energy density in UniKEF not only creates gravity and spatial dimension (distance) but that the flow of energy causes change and change is the root of the concept of time. Diminishing the energy density in the production of gravity contracts distance or curves space and reduces energy flow or changes apparent time flow.

The difference is in viewing time as a 4th dimension versus being an attribute of an energetic 3D space.

The volume of space that is in between masses repels them. The external space in the source cone pushes masses together. The zero degrees angle penetration rays, (See Fig 26) where the trigonometric function = 1.0000, does not diminish with distance.

Hence the UniKEF function trends toward unity at infinity and not toward zero as the Newtonian view do.

That feature causes UniKEF to have stronger gravity at galactic distance scales but becomes repulsive at cosmic scales where there is greater distance separating objects than there is external distance to the finite edge of the universe.

The energy density along the line of gravity establishes the spatial dimension (distance) between bodies in terms of "Quantitative" energy. As gravity is produced by attenuation, energy density diminishes and hence distance is reduced. Gravity causes curved space not curved space causes gravity.

The distance between objects varies as a function of the mass of the observer and the observed. This affect is minor over short ranges and for moderate gravitational masses.

This affect may be showing up in the (4) deep space probes velocity anomaly observed in recent years.

These probes appear to be traversing a different distance than we calculate they should by them being outside our line of gravity to the sun and other solar masses to them. See Fig's 21a & b.

Refraction:

Refraction is the apparent slowing down of light passing through various materials. In reality, light maintains its normal speed but its velocity (direction of motion) is continuously changing and causes it to travel greater distance through the material than we measure as remote observers.

From light's perspective, it travels at the same speed but goes further than we see it travel.

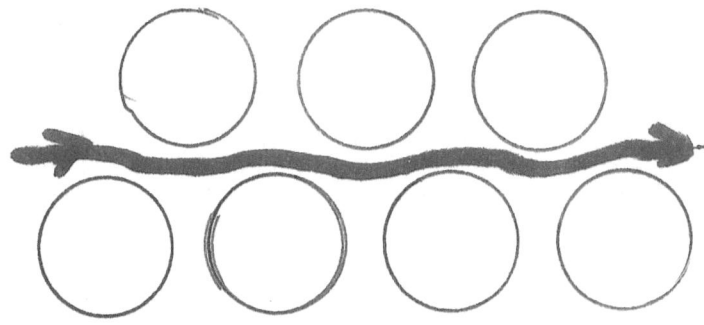

Low Index of Refraction

Fig #31a

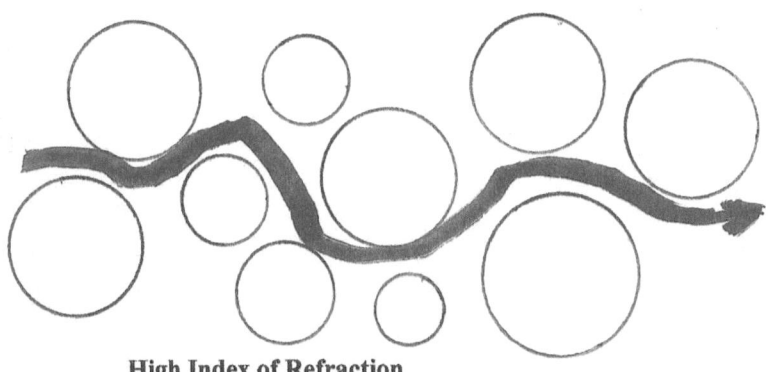

High Index of Refraction

High Index of Refraction
Fig #31b

Recall that in UniKEF mass is nothing more than compacted space and space is flowing energy. The compaction factor is believed to be associated with the mass to energy ratio according to $E = mc^2$.

Light is not traversing the compacted space geometry, which is involved in gravity production; it is traversing along the outer surfaces of the compacted space, which are curved.

The amount of deviation in a vector path through which light must traverse determines the index of refraction. Our measure

Dan Keith McCoin

of light speed into and out of a material does not see the zigzag path that light may follow.

A modern view is that photons are absorbed by the atom and re-emitted after some energetic decay process. This causes light to travel $v = c$ in intermittent spurts and results in an apparent slowing of light even though light may be propagation at $v = c$ while in motion.

CHAPTER 11

Testing UniKEF

The gravity testing that has been done needs to be replicated to either confirm or falsify the original data.

Other tests may also prove interesting. For example, test two flat plates in a variety of positions.

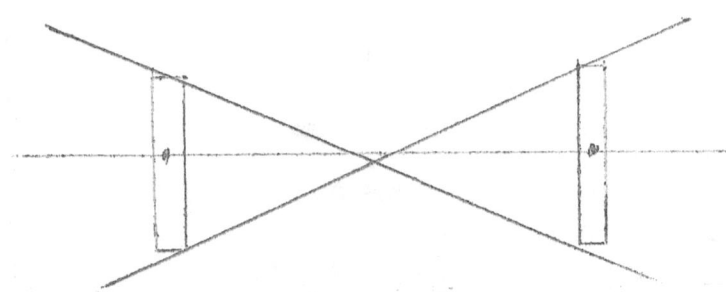

Both vertical

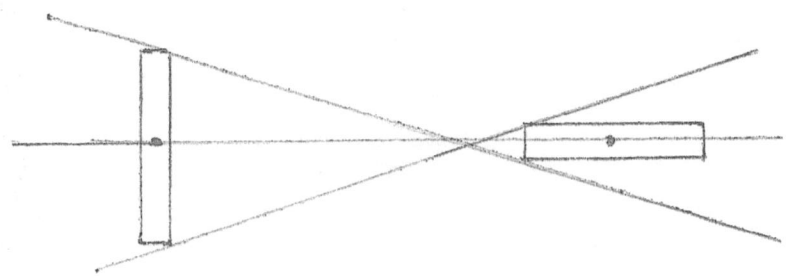

One vertical one horizontal

Both Horizontal
Fig 32

Cone of Sources & Geometry Affects

The only thing changing in this test is the UniKEF COS (Cone of Sources). Mass, center of mass and distance between centers remains constant.

Other geometric objects such as cubes, cones and sphere combinations should show a geometric affect even when they have the same mass and are tested at the same separation distance.

Another test I suggest is to check for dynamic type ether masking an absolute type system of relativity. In space test to see if, measurement of absolute velocity can be achieved.

While in a gravity field testing indicates that light carries forward momentum of the source such that a pulse of light projected orthogonally off a moving platform the pulse beam moves concurrent with the platform. Aberration of star light suggests light in deep space may propagate differently.

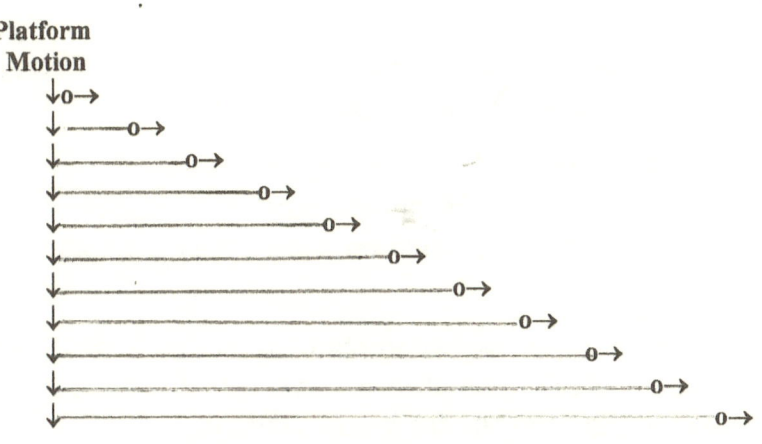

Fig 32 Photon Forward Momentum

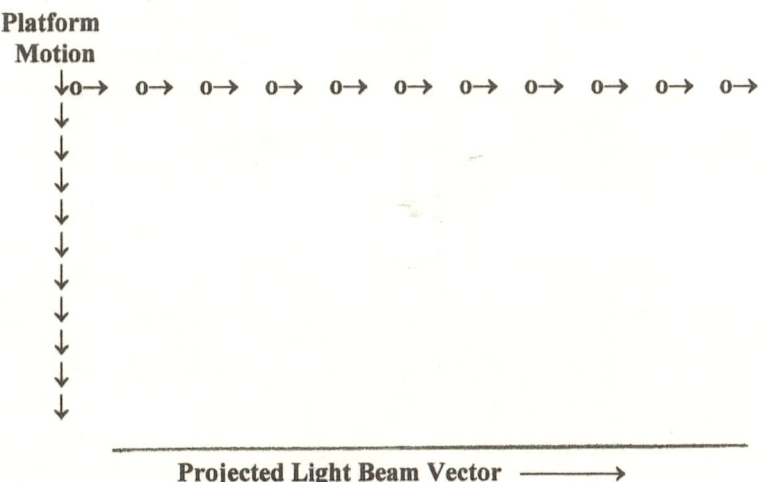

Fig 33 Photon No Forward Momentum

Since the photon purportedly has no mass then it should carry no forward momentum and follow the path of Fig 33.

If however, the photon has any immeasurable mass or the relativistic mass of the photon created carries the forward momentum of the platform then the test will not produce a means of detecting absolute motion.

If it follows the path of no forward frame momentum, it will prove the concept of an absolute rest reference frame. A projected light beam would be such a frame if it carries no forward momentum.

This test may be attempted using a Bose-Einstein Condensate (BEC) Chamber to slow light to ranges of test velocity. i.e. 360km/sec for galactic motion.

The device consists of a Laser, a BEC Chamber, a Chromatic Filter, a Chromatic Analyzer and Computer.

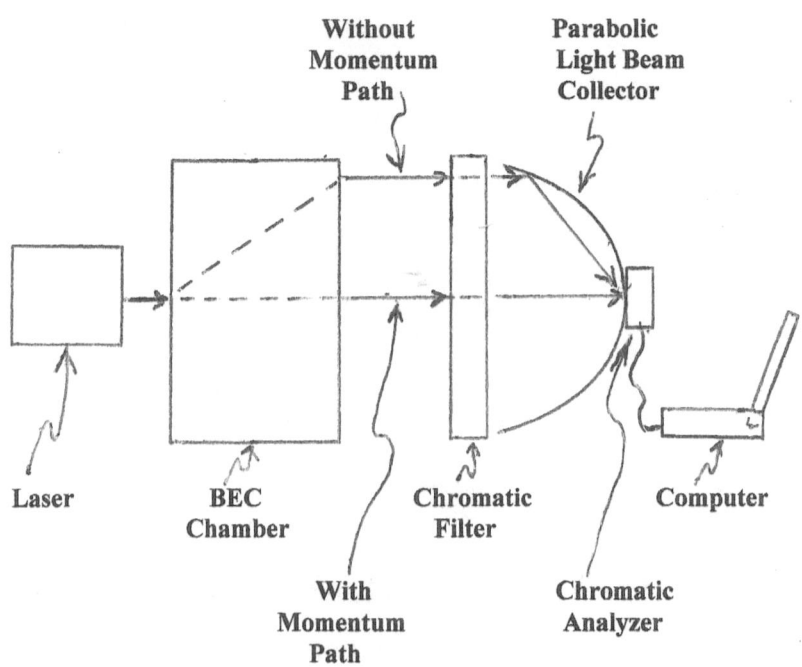

Fig 34

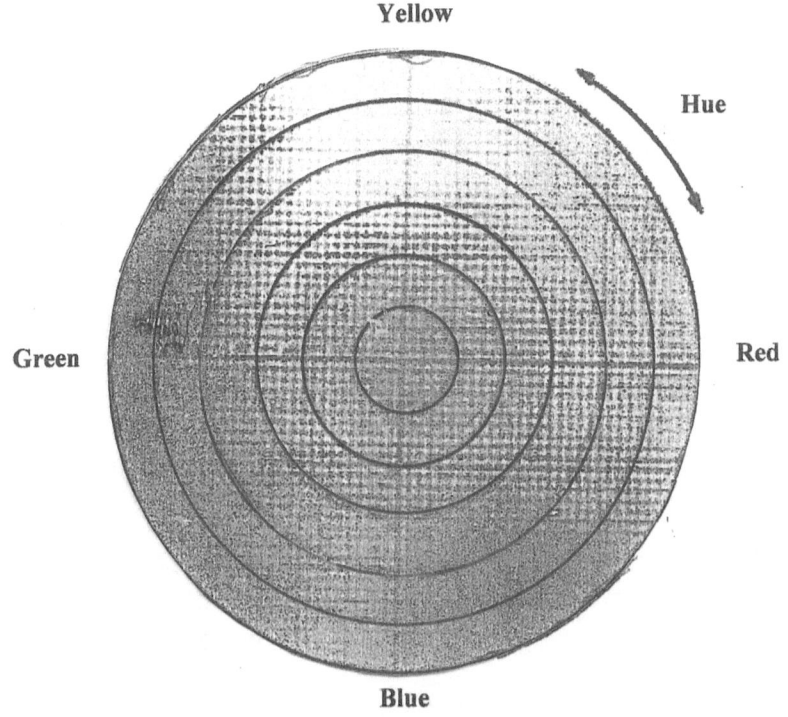

Chromatic Filter
Fig 35

It can be seen that if light is slowed down to 360km/sec through the BEC and the only motion of the platform is 360km/sec then a shift in where the beam passes through the chromatic filter determines the direction and magnitude of the motion, IF there is no forward momentum in the projected light beam.

The BEC needs to be tunable or interchangeable with other BEC chambers to have a series of velocity ranges to test.

The computer detects the hue and color of the light and computes the platform vector and velocity based on the BEC light velocity and length of travel through the BEC.

Since relativistic affects are exponential with velocity, perhaps particle decay tests in an accelerator, which is properly aligned, could enhance computation of our possible absolute motion in the universe. Taking c as equal to 3E3 m/s:

Assuming 360 m/s cosmic motion only yields 1 second of time loss every 44,012 years.   I suggest they combine that velocity with a particle beam accelerated to perhaps 0.999c.

Then run multiple tests during the year and at two times of day when the beam has a tangential vector to earth's orbit due to earths rotation, so as to provide both %c + v and %c - v between the currently measured cosmic velocity to the CMB (Cosmic Microwave Background) and the particle beam.

The cosmic motion is at 10° from the solar elliptical plane such that any measured deviation should be 98.48% representative of actual cosmic motion.

Doing tests in this manner should produce a high and low time dilation calculation over the period of a year and be indicative of both magnitude and vector of motion.

i.e.:   Where the beam accelerator is located at earth's equator rotational velocity = 463.8 m/s.

V1 = 0.999c + orbit + rotation tangent to orbit + cosmic motion = 2.997E8 m/s +29,780 m/s + 463.8 m/s + 360 m/s = 299,730,603.8 m/s. Divided by c = 0.999102013c that yields a gamma of 23.6019288. Time dilation is 1 hour, 1 minute and 43.07 seconds/day.

V2 = 0.999c - orbit - rotation tangent to orbit - cosmic motion = 2.997E8 m/s - 29780 m/s -

463.8 m/s - 360 m/s = 299,669,396.2 m/s. Divided by c = 0.998897987c yielding a gamma of 21.306464453 which induces a time loss of 1 hour, 7 minutes and 35.1 seconds/day

Such high velocity testing incorporating low velocity contributions placed in co-moving and counter moving alignments yields substantial differentials making it far more precise to identify any possible absolute motion in the universe.

Instead of merely trying to measure 1 second in 44,012 years, you can now measure a change of 5 minutes and 52.03 seconds/day.

Locating an absolute universal rest reference, should it exist, is easy using today's technology if they have any interest in actually learning the truth rather than merely protecting Einstein's view.

Finally a test to determine the speed of gravity.

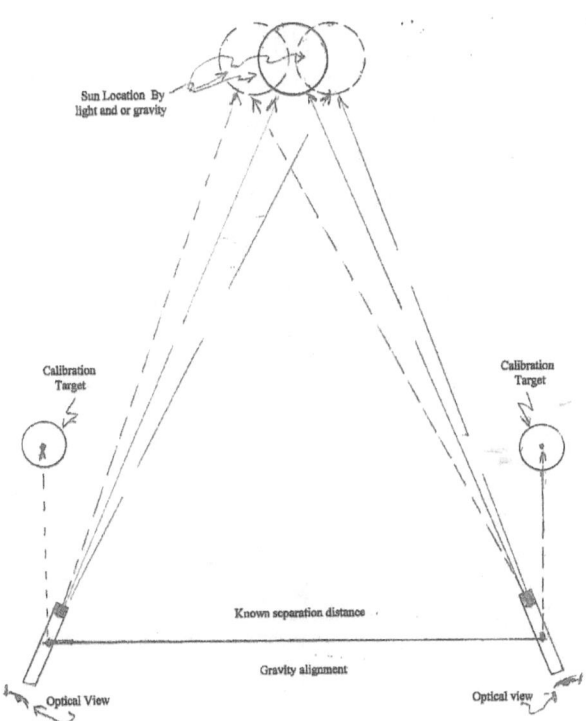

Fig 36:   Test for Speed of Gravity.

A pair of matched optical tubes with a means to force rotational motion when desired to optically center the sun in the lens and A Cavendish Balance component left to orient to

Dan Keith McCoin

the sun by gravity and equipped with laser calibration targets to match angular displacement graduations between the devices.

The instruments need micro-angular displacement graduation adjustments to allow proper alignment of optical components and gravity components and to measure any deviation between speed of gravity and light velocity when allowed to focus on the Sun.

It will be necessary to run time responses on the Cavendish Balance instruments to determine their response times to gravitational force changes and compute the normal delay in alignment to actual gravity pull of the Sun.

In this arrangement should the optics lag the gravity position then we know gravity travels faster than light. UniKEF predicts gravity will be found to be virtually instantaneous such that gravity should be in advance of light once proper delay is determined for the gravity balance.

If they are both focused on the same location then gravity and light has equal velocity. Should light be advanced then gravity moves slower than light.

130

DEPARTMENT OF THE ARMY
HEADQUARTERS UNITED STATES ARMY MATERIEL COMMAND
WASHINGTON, D.C. 20315

AMCRD-PS-P
CN 686-69

2 6 APR 1978

Specialist Six Dan K. McCoin
Post Office Box 33
Fort Davis, Canal Zone   09829

Dear Specialist McCoin:

This is in response to your letters requesting an AMC evaluation of the
technical material you submitted concerning UniKEF. Your submissions
have been carefully evaluated by personnel of the Research, Development
and Engineering Directorate of this Headquarters.

From querying both physicists and mathematicians, we found out that much
of the substance of your ideas regarding gravitational absorption has appeared in existing literature. We have at hand a German publication,
"Observation of the Eclipse of 1954 in Norway," which contains material
quite similar to your theory, although further advanced. The document is
rather lengthy but a copy will be forwarded under separate cover following translation.

Your work displays a considerable amount of ambitious and serious thought.
Those evaluating it encourage you to continue, although AMC cannot support
any experimentation in this area at this time.

It is suggested that additional formal education would be of decided advantage in the pursuit of your efforts. Toward this end, we are inclosing
a copy of AR 350-200 for your information. Also attached is a copy of
DA Pamphlet 360-613 which outlines educational opportunities for Army
personnel. Your submissions are being returned for further use. Copies
will be retained by this office.

Although it is regrettable that a more favorable response cannot be made,
your interest in bringing your work to the attention of this Command is
very much appreciated.

Sincerely yours,

6 Incls
as

ROBERT B. MERCER
Lieutenant Colonel, GS
Acting Chief, Project Plans
Research, Development & Engineering Dir

ISTITUT FÜR ANGEWANDTE GEODÄSIE
— Der Direktor —

4200/71

6 Frankfurt a. M., den 7 April 1971
Kennedyallee 151
Telefon (0611) 61 04 91
Hausanschluß:

Mr. Dan K. McCoin

Box 5003

Ft. Amador, C.Z., 09834
USA

Subject: Reprint of an extract from IfAG Mitt. No. 16,
DGK Reihe B No. 234

Dear Sir,

Thank you for your letter of 22 December 1970, which we have
received, together with a copy of your manuscript, only recently,
due to an incomplete address.

In reply to your letter we beg to inform you that we make no
objections to publish an extract from the report "Beobachtungen
zur Sonnenfinsternis 1954 in Südnorwegen" by Dr. Brein, Jelstrup,
Nottarp, Sandig and Sigl, as taken from Mitt. No. 16 and DGK-
Veröffentlichung Reihe B No. 34, respectively.

We agree to the publication in the form you have presented to
us, and we herewith give you our permission to reprint the
extract as indicated.

Please find enclosed the copy of your manuscript.

Sincerely yours,

H. Knorr

Prof. Dr.-Ing. H. Knorr
Director

Enclosure

132

www.ingramcontent.com/pod-product-compliance
Lightning Source LLC
Chambersburg PA
CBHW030007190526
45157CB00014B/959